Gangetic Fish Otoliths

THE AUTHORS

Dr. Salman Khan is currently working as an Assistant Professor, School of Life Science and Technology, IIMT University, Meerut. He was awarded his Ph.D. degree under the supervision of Professor M. Afzal Khan, Laboratory of Fish Biology and Otolith Research, Department of Zoology, Aligarh Muslim University, Aligarh. He is a dedicated researcher with a good number of national and international publications. He has participated in various national as well as international Conferences/Symposia/Workshops and received Best Paper Award for his presentations thrice at different conferences. One of his recent otolith research appeared in Scientific Reports which belongs to the reputed "Nature group of journals". He is not only an excellent researcher but a very good communicator of science. He recently won first prize in an international competition in Australia and was conferred the "Science Communication Award-2017" at a special conference organized by Australian Society of Fish Biology at Albany, Australia. Throughout his research career, Salman has received funding from CST-UP. as Research Assistant, UGC Non-NET fellowship and CSIR-Direct SRF. Salman's research focuses mainly on developing the precise and scientifically sound protocols for age and growth estimation, and stock delineation in commercially important freshwater fishes. He shows great deal of interest in exploring otoliths and working on its applications in devising the strategies for management, diversity and conservation of natural fish populations.

Dr. M. Afzal Khan is currently working as a Professor at Department of Zoology, Aligarh Muslim University, Aligarh. Prof. Khan is heading a very dynamic research group working on biology of freshwater teleosts, their population dynamics and natural resource management. His notable research contributions are on identifying species-specific structure(s) for precise fish age estimation; to decipher species-specific growth parameters, and to delineate stock(s) of freshwater teleosts mainly from the Gangetic river system. Prof. Khan is known for his pioneering research contributions in India on precise age estimation in fishes and use of otoliths for fish stock delineation. He has also worked on fossil otoliths to decipher the probable biological information at Ludwig Maximilians University, Munich, Germany. He along with one of his students and first author of this book, Salman Khan, has developed an indigenous, efficient and highly economical "Otolith polisher" that is currently installed at his laboratory. Prof. Khan's laboratory received research projects from various funding agencies such as UGC, DST-SERB and CST-UP. His research findings appear mostly in SCI indexed and impact factor journals published from Nature Publishing Group, Elsevier Science, Wiley-Blackwell, Blackwell-Verlag, Wiley Inc., CSIRO publishing, EDP Sciences, Taylor and Francis, Springer, *etc.*

Gangetic Fish Otoliths

— Authors —

Salman Khan

Laboratory of Fish Biology and Otolith Research,
Department of Zoology, Aligarh Muslim University,
Aligarh-202002, U.P., India

Currently at: School of Life Science and Technology,
IIMT University, Meerut–250001, U.P., India

Mohammad Afzal Khan

Laboratory of Fish Biology and Otolith Research,
Department of Zoology, Aligarh Muslim University,
Aligarh-202002, *U.P., India*

2022

Daya Publishing House®
A Division of
Astral International Pvt. Ltd.
New Delhi – 110 002

ISBN: 9789354614767

Publisher's Note:

Every possible effort has been made to ensure that the information contained in this book is accurate at the time of going to press, and the publisher and author cannot accept responsibility for any errors or omissions, however caused. No responsibility for loss or damage occasioned to any person acting, or refraining from action, as a result of the material in this publication can be accepted by the editor, the publisher or the author. The Publisher is not associated with any product or vendor mentioned in the book. The contents of this work are intended to further general scientific research, understanding and discussion only. Readers should consult with a specialist where appropriate.

Every effort has been made to trace the owners of copyright material used in this book, if any. The author and the publisher will be grateful for any omission brought to their notice for acknowledgment in the future editions of the book.

Published by : **Daya Publishing House®**
A Division of
Astral International Pvt. Ltd.
– ISO 9001:2015 Certified Company –
4736/23, Ansari Road, Darya Ganj
New Delhi-110 002
Ph. 011-43549197, 8130496929
E-mail: info@astralint.com
Website: www.astralint.com

DEDICATED TO

Late Prof. Syed Zahoor Qasim

Padma Bhushan and Padma Shri
Leader of First Indian Antarctic Expedition
Former Member, Planning Commission of India

Former Vice-Chancellor,
Jamia Millia Islamia University,
New Delhi

Prof. A. K. Jafri

Formerly,
Chairman, Department of Zoology
Dean, Faculty of Life Sciences, and
Dean Students Welfare
Aligarh Muslim University, Aligarh

Acknowledgements

We are thankful to the Chairman, Department of Zoology, A.M.U., Aligarh for providing the laboratory facilities related to my work. Most cordial thanks are due to Dr. Shahista Khan, Dr. Kaish Miyan, and Ms. Ankita for their manifold services, advice, and helping hands offered at various junctures of this work.

We would like to acknowledge the constant support provided by Mr. Sam Khan, Department of Mass Communication, Aligarh Muslim University, Aligarh for his kind help in making the fish images.

Salman Khan is grateful for the financial assistance provided by Council of Science and Technology, U.P., Lucknow (as Research Assistantship), University Grants Commission, New Delhi (Non-NET fellowship) and Council of Scientific and Industrial Research, New Delhi, (as Direct-Senior Research Fellowship financial assistance).

Salman Khan

Mohammad Afzal Khan

ICAR-Directorate of Coldwater Fisheries Research
Bhimtal-263136, Uttarakhand

Foreword

Otoliths are sometimes referred to as "Biological CD-ROMS" owing to their unique property of being metabolically inert, growing throughout fish life and retaining the chemical signatures of the environment (on a daily/sub-daily basis) in which fish has lived. Fish otoliths come out undigested along with the faecal matter of piscivorous animals. Also, the shape of otoliths is known to show intraspecific variation with respect to time and space. Thus, otoliths are used worldwide to infer information on fish's life history, habitat utilization, natal habitat, larval duration, feeding habitat, breeding habitat and migratory pattern. India is fortunate to have huge natural fishery resources both marine and freshwater, being the 3rd largest country in fish production and 2nd largest in aquaculture production in the world with a record production of 142 lakh tons in 2019-20. Unfortunately, the ichthyofaunal diversity in particular and aquatic diversity in general, and their conservation strategies are not as successful as they should be mainly owing to the lack of updated, sufficient and scientifically reliable data. It is the need of the hour to work more intensively on various aspects of fish biology specially otoliths in order to appropriately address the research problems discussed above.

It is indeed a commendable effort by the authors to bring out this immensely informative and beautifully scripted book, which is appropriately and engagingly supported by clear and self-explanatory diagrams and photographs. This book consists of photographs and descriptions of the otoliths of 51 species belonging to 17 families *viz.*, Channidae, Gobiidae, Nandidae, Ambassidae, Schilbeidae, Clariidae, Sisoridae, Heteropneustidae, Bagridae, Siluridae, Cobitidae, Mastacembelidae, Belonidae, Clupeidae, Notopteridae, Mugilidae, Cyprinidae, which are distributed in eight orders.

I congratulate Prof. M. Afzal Khan and his research team for bringing out a very informative and content-oriented publication which will be of great help for graduate, post-graduate, Ph.D. students of fisheries, zoology, animal sciences and multiple stake holders. I believe that the students, academicians and researchers will make full use of the interesting information in the book in order to deal with the pertinent and contemporarily relevant research problems.

Dr. Debajit Sarma

Director

ICAR-DCFR Bhimtal

Preface

Otoliths, the fish ear bones, are described in this book. All bony fishes have three pairs of otoliths: the sagitta, asteriscus, and lapillus. These otoliths are made up of calcium carbonate, in the form of aragonite, in a protein matrix. They are present within membranous labyrinths in paired otic capsules on either side of the skull.

Otoliths are unique among hard anatomical structures of fish because they are metabolically inert. Fish otoliths can tolerate the digestive juice action in the predator's gut. Therefore, they are useful scientific tools to study the food and feeding habits of piscivores especially in the circumstances when other fish bones and scales have been already destroyed. Thus, otolith science and research are important not only to academics but have immense important applications in understanding the population structure, precise age estimation and growth assessment, identifying stocks that need to be managed separately, studying the natal habitat, movement patterns and breeding ground in a fish life cycle, and also helping to identify the food of piscivoers. The importance of fish otoliths are not only limited to ichthyology and fisheries, but are extensively used to the study of the feeding ecology of fish predators, and various aspects of palaeontology, archaeology, stratigraphy and zoogeography. This book consists of photographs and descriptions of the otoliths of 51 species belonging to 17 families *viz.*, Channidae, Gobiidae, Nandidae, Ambassidae, Schilbeidae, Clariidae, Sisoridae, Heteropneustidae, Bagridae, Siluridae, Cobitidae, Mastacembelidae, Belonidae, Clupeidae, Notopteridae, Mugilidae, Cyprinidae, which are distributed in eight orders.

The shape and size of otoliths is not only species-specific but intraspecific variations are also noted in the fish species distributed in different habitats. In the present book, a brief description of fish, its classification, specimen photograph and a preliminary account of its morphology are given. We have selected the largest otolith (amongst lapillus, sagitta and asteriscus) in each fish species for description. Currently, otolith research is progressing at a rapid pace around the globe and so India will also not be behind. The biggest hindrance at the moment is that we are not having enough published information on fish otoliths from several ecologically and commercially important aquatic resources. Since long, it was warranted to have such book on the fish fauna of freshwater ecosystems. This book is just a beginning of compiling such useful information. Comprehensive studies to describe the otoliths of extant as well as extinct ichthyofauna of the Indian sub-continent are highly warranted.

Salman Khan

Mohammad Afzal Khan

Contents

Chapter 1

Introduction

Otolith or ear stones present in the inner ear of fishes are responsible for sound detection and maintenance of balance (Payan *et al.,* 2004). The three pairs of otoliths, lapilli, asterisci and sagittae, are lodged in, utricle, lagena and saccule, respectively of the inner ear in the teleost. Otoliths mainly consist of calcium carbonate ($CaCO_3$) structures, which crystallise around an organic matrix with daily periodicity (Campana 1999). The organic otolith component constitutes between 2 and 10 per cent of the total otolith mass and primarily comprises of proteins including lipoproteins and glycoproteins, glycosaminoglycans and polysaccharides; 50 per cent of these are water soluble (Kalish 1991; Campana 1999; Dauphin and Dufour 2003). Each otolith pair is different in shape and appearance, and in most adult fishes the sagittae are larger and have the largest morphological variability (Campana 2004; Tuset *et al.,* 2008). To fisheries biologists, the otolith is one of the most significant tools for understanding the life of fish and fish populations; growth rings (annuli) similar to those of a tree record the age and growth of a fish from the date of hatch to the time of death (Casselman 1987). There are numerous published reports on precise age estimation and growth assessment using the growth rings on otoliths in marine fishes (Karen *et al.,* 2003; Sipe and Chittenden 2001) as well as freshwater fishes (Khan *et al.,* 2009; Khan *et al.,* 2017). Moreover, chemical and elemental assays allow the reconstruction of significant life history events, such as the year of hatching, migration pathways, population identity, and the temperature of the water (Campana 1999). Otoliths also act as a chemical tag by recording and retaining the elemental profile of the waters inhabited by the fish throughout its life (Elsdon *et al.,* 2008; Khan *et al.,* 2012; Miyan *et al.,* 2016). Otolith microchemistry has been used in retrospective examinations of diadromous fish life history due to the ability of the technique to detect changes in salinity experienced by individuals, largely through changes in the Sr concentration (Limburg 1995; Secor *et al.*, 1995; Limburg 1998; Secor and Rooker 2000; Zimmerman 2005; Diouf *et al.*, 2006). The characteristics, shape and growth pattern of otoliths are highly species specific (Frost 1981; Hunt 1992; Aguirre and Lombarte 1999). Low degradability of otoltihs while passing through the gut of the ichthyophagous species makes them an important tool for studies on stomach contents (Cottrell *et al.*, 1996; Tollit *et al.*, 1997; Everson *et al.*, 1999; Granadeiro *et al.*, 2000; Gomiero and Braga 2007). In the last decade, several studies have shown that otoliths provide a clear phylogenetic signal and some of their morphological characteristics were used in phylogenetic studies (*e.g.*, Lombarte *et al.*, 2010). In most fishes, the largest otolith is the sagitta, while the lapillus and astericus are very small. However, In the members of Cypriniformes and Siluriformes, the asterisci are the largest otoliths followed by lapilli and sagittae (Harvey *et al.*, 2000; Campana 2004). In this book, the word 'otolith' generally indicates the sagittae, while the lapilli or the asterisci will be individually noted. Otoliths are also reported to be metabolically inert and thus do not reflect physiological changes that might occur throughout the life of the fish (Phelps *et al.*, 2007). Otoliths are also more useful than opercular bones, spines, scales or any other fish remains in generating different biological or ecological information. The information on fish otoliths have been published since late 1970s (Wirtz 1976; Nolf 1979; Morrow 1977) and continuing till date for example Rousset 1983; Akkiran 1985; Abilhoa and Corrêa 1993; Hofstaetter *et al.*, 1996; Volpedo 2000; Assis 2000; Campana 2004; Furlani *et al.*, 2007; Artzi *et al.*, 2009; and Bremm and Schulz 2014. Many of such studies are quite extensive and specific, thus being used in species identification for example, Assis 2005; Tuset *et al.*, 2006 and 2008; Lombarte and Cruz 2007; Chulin and Chen 2013; Rojo 2015 and Sánchez and Martínez 2017. The AFORO (Shape Analysis of Fish Otoliths) database has been recently established that offers an interactive catalogue including an automatic system of species identification based on the mathematical description of the otolith characteristics (Lombarte *et al.*, 2006). However, from India, there are very few published reports showing a compilation of fish otoliths' description. Rana (1988) described 12 fish species based on otoliths recovered from the Deccan Trap associated sedimentary (Intertrappean) beds exposed near Rangapur, Hyderabad District, Andhra Pradesh, India. In a study on Marine Middle Eocene fish otoliths from India and Java revealed 15 and 24 teleost taxa, respectively (Nolf and Bajpai 1992). Kumar *et al.* (2012) described the otolith morphology of 12 species of sciaenids occurring along the north-west coast of India. Therefore, on account of paucity of published reports on fish otoliths from Indian waters and no such information on Gangetic fishes, the present study was undertaken to provide a photographic guide of the otoliths of the most common fish species from the Ganga river. The book also contains description on otolith and also a brief note on habitat, distribution, general characters and conservation status of the fishes.

Otolith Location and Function

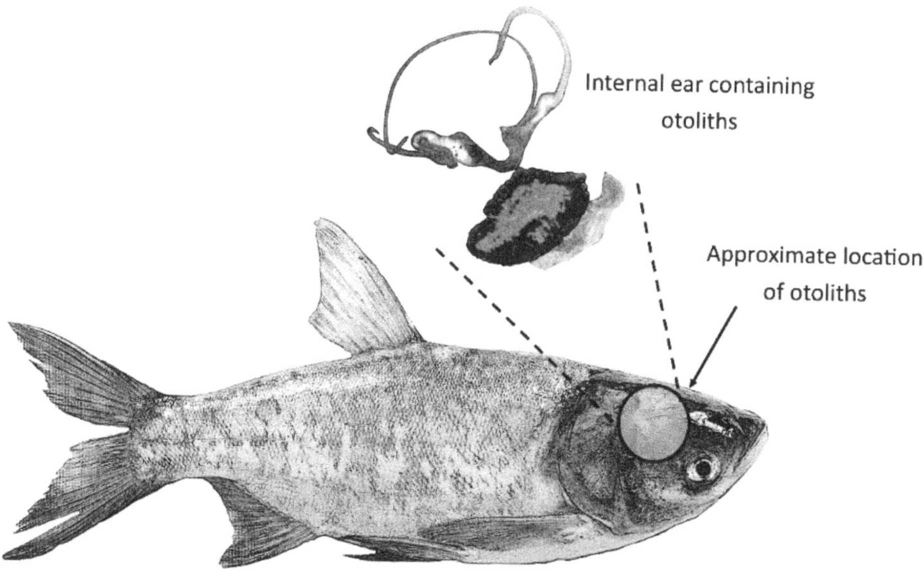

Figure 1 (a): Location of Internal Ear Housing the Otoliths in Fish.

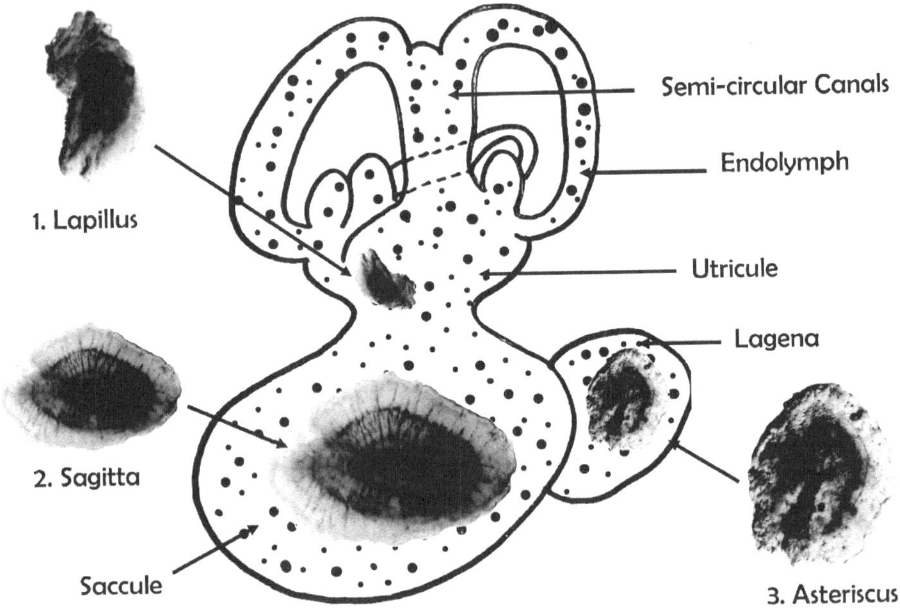

Figure 1(b): Types of Otoliths and their Location in the Internal Ear of Fish (Modified from Panfili *et al.*, 2002). Otolith images are of *C. punctata* (Khan *et al.*, 2019).

Otolith Composition

On the basis of a comprehensive suite of assays, Campana *et al.* (1997) reported that otoliths of *Micropogonias ungulatus* composed of 96.2 per cent by weight of calcium carbonate, with an additional 0.73 per cent present as non-organic trace

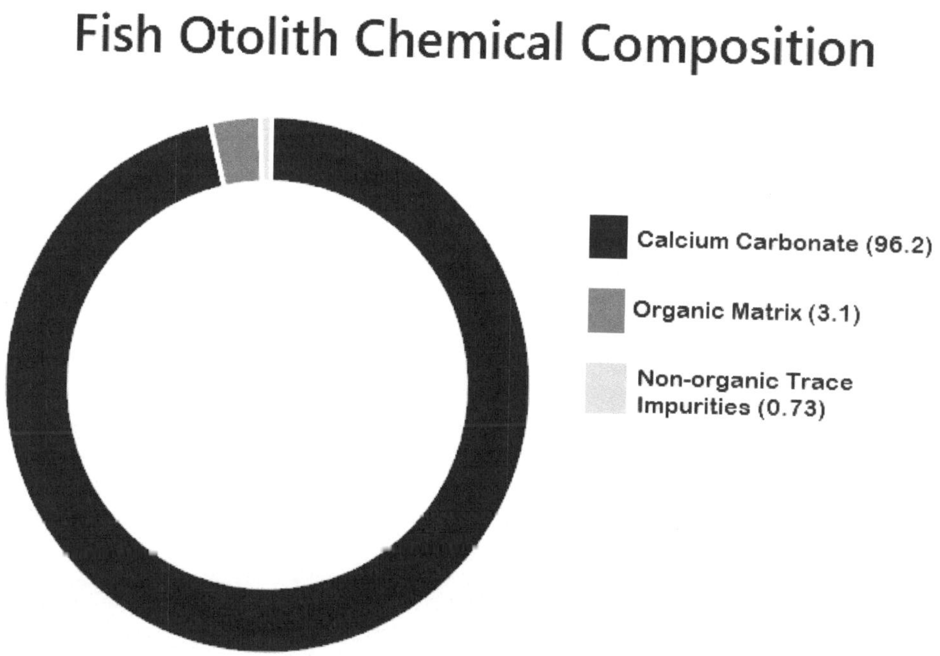

Figure 2: Showing Chemical Composition of a Fish Otolith (Based on the otolith composition reported by Campana *et al.*, 1997).

impurities. The remaining 3.1 per cent represents the organic matrix. The elemental composition is dominated by the major elements calcium, oxygen and carbon, which make up the calcium carbonate ($CaCO_3$) matrix. However, the majority of the elements are present at the minor (> 100 ppm) and trace (< 100 ppm) levels. The minor elements are represented by Na, Sr, K, S, N, Cl and P, while the bulk of the trace elements are present at concentrations of less than about 10 ppm.

Otolith Morphology

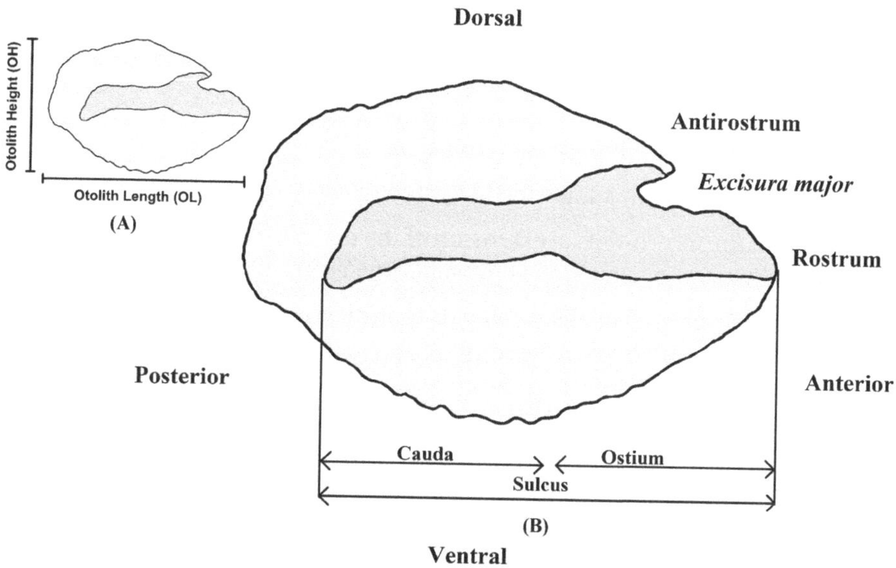

Figure 3: (A) Otolith Length and Otolith Height Measurements; (B) Medial Face of Otolith Showing the Features.

Definition and Description about Otolith Morphology

The definitions and descriptions used to describe the otoliths in this study are as under:

Dorsal area: The region lying dorsal to sulcus.

Ventral area: The region lying ventral to sulcus.

Antirostrum: An antero-dorsal extension of the otolith, above notch. Antirostrum is absent when no extension is present or excisura major is absent.

Cauda: Posterior part of sulcus, posterior to neck constriction.

Ostium: Anterior part of sulcus, anterior to neck constriction.

Sulcus: Depression on the medial face, usually longitudinal and located in the center. In most cases, it can be divided into ostium and cauda.

Rostrum: An anterior extension of the otolith, measured from the posterior end of notch to anterior margin of the otolith. Rostrum is absent when no extension is present or excisura major is absent.

Excisura (or excisura major): Opening of the sulcus on the anterior margin. If sulcus does not reach to the margin, there is no excisura.

Chapter 2

Materials and Methods

Fish samples were collected during the period November 2012 to December 2019 from the River Ganga and two of its tributaries, the River Gomti and the River Yamuna in India (Fig. 4). Fishes were collected using commonly employed fishing gears, such as, cast net, drag net and gill net. Following the keys and descriptions given by FishBase and/or Jayaram (1999), the samples were identified and their general characters, distribution and habitat provided. The conservation status of the selected fishes was confirmed from the IUCN website. Total length (TL) of the fish was measured to the nearest mm. To extract the otoliths, the supraoccipital and frontal bones were cut, the brain removed and the inner ear exposed under a stereozoom microscope (Nikon SMZ745T). Subsequently, the otoliths from both the sides of the head were extracted, obtaining the vesicles and corresponding otoliths (Sánchez and Martínez 2017). Otolith length and height were measured as per the description given in Lin and Chang (2012). Otolith lengths (OL) were measured to the nearest 0.01 mm as the greatest distance from the anterior most point to the posterior most margin, approximately parallel to the sulcus. Otolith height (OH) was recorded to the nearest 0.01 mm as the greatest distance between the dorsal margin and the ventral margin, perpendicular to OL. The photographs were taken with a Nikon Stereozoom Microscope attached to a video camera, connected to a computer. Otolith morphology was described according to Smale *et al.* (1995).

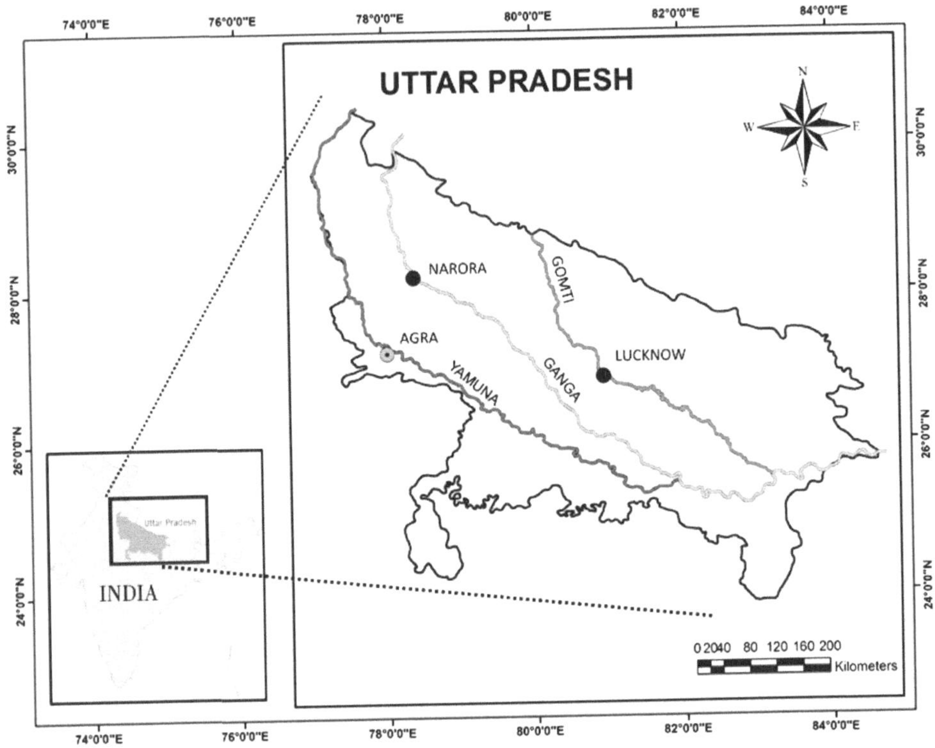

Figure 4: Map Indicating the Three Sampling Sites.

Chapter 3

Fish and Otolith Description

The description and the photographs of fishes and their otoliths have been given as per the family and the order they belong.

Order: PERCIFORMES

Family: Channidae (Snakeheads) Plate 1 and 1a

Species sampled	*Channa marulius, Channa punctata, Channa striata, Channa gachua*
Collection of samples	Gomti River, site: Lucknow (26.8467° N, 80.9462° E)
Habitat and Distribution (FishBase)	Habitat preference varies by species, most occupy streams and rivers. Snakeheads can also be found in swamps, rice paddies, ponds, reservoirs, ditches, and canals. These species are distributed in the freshwaters of Asia region: India to China, Afghanistan, Pakistan, Sri Lanka, Nepal, Bangladesh, Myanmar and also in the Thailand and Cambodia region.
General Characters (Jayaram 1999)	Body elongate, cylindrical anteriorly. Body with large scales, head with plate-like scales. Cephalic pits present. Teeth on jaws, vomer and palate. Branchiostegal rays five. Gill membranes connected beneath isthmus. A suprabranchial organ present. A single long spineless dorsal fin present and a similar anal fin, both free from caudal fin. Lateral line abruptly curved or almost interrupted. Air bladder present, not bifid posteriorly.

Conservation status (IUCN)	All the species were in the "Least Concern" category
Otolith morphology	Otolith shape: approximately oval, Thickness: moderately thick, Ventral margin: rounded and slightly irregular, Posterior margin: sharply rounded and irregular, Rostrum: small and sharply rounded to rounded in large otoliths, Antirostrum: Small and sharply rounded, Excisura: narrow to moderate and slit-like, notch very shallow, angle acute.

Family-Channidae

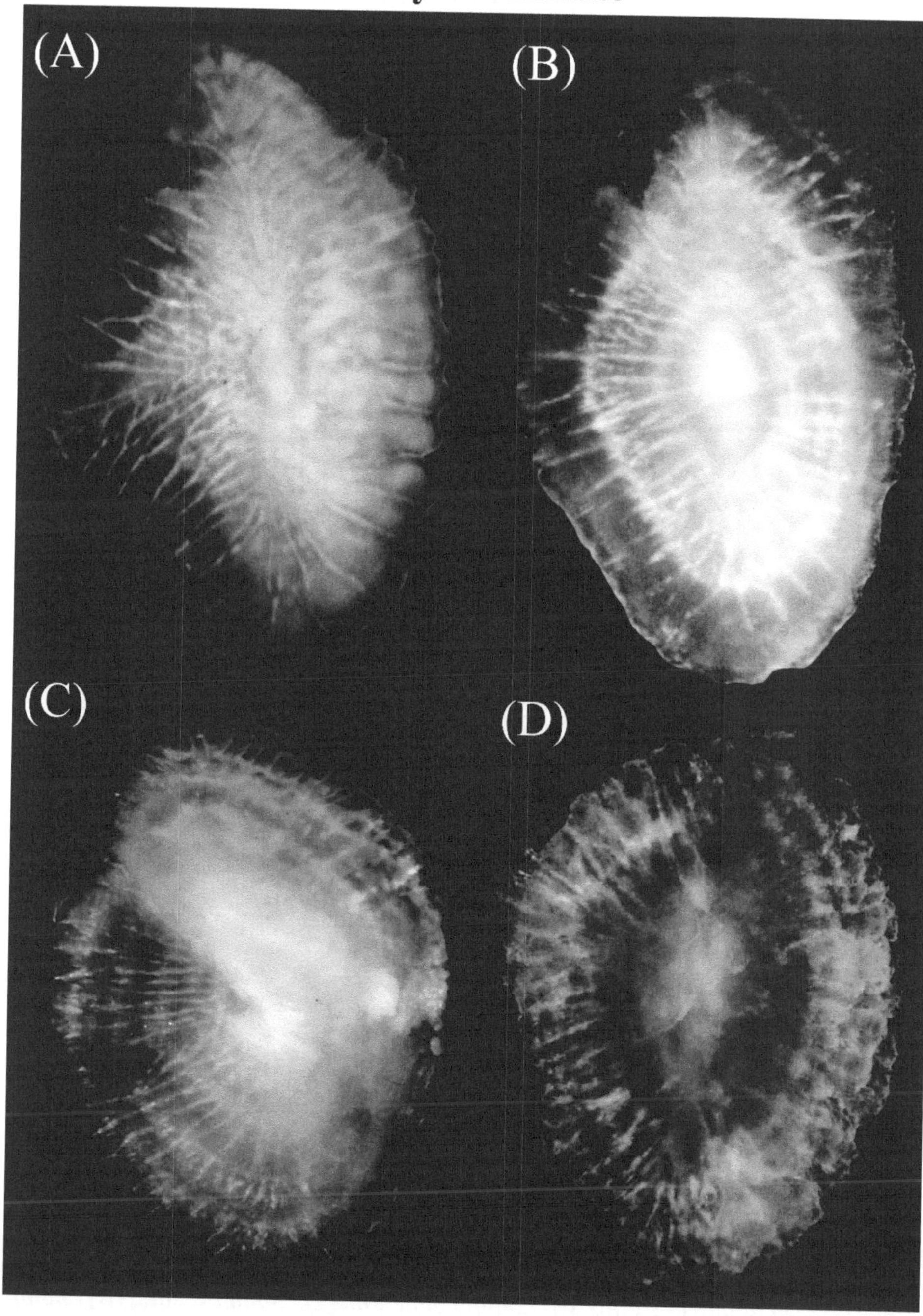

Plate 1: Otolith (sagitta) of (A) *Channa marulius* (OL=8.94 mm; OH=6.78 mm), (B) *Channa punctata* (OL=6.99 mm; OH=4.22 mm), (C) *Channa striata* (OL=14.13 mm; OH=7.39 mm), *Channa gachua* (OL=4.1 mm; OH=3.2 mm) collected form the river Gomti.

Family– Channidae

(A)

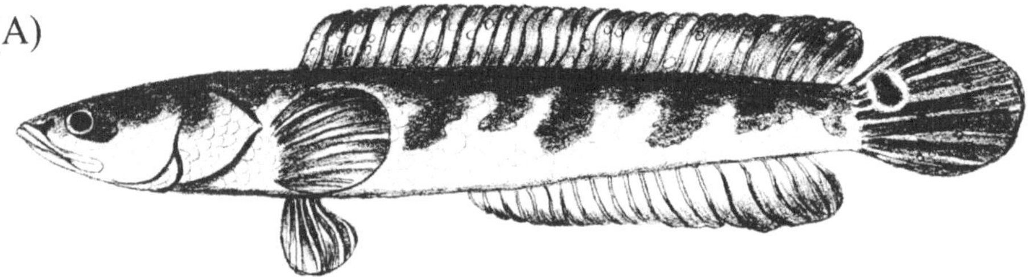

(B)

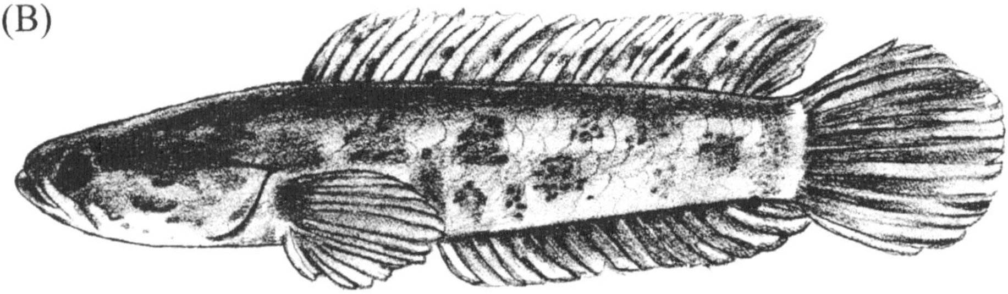

(C)

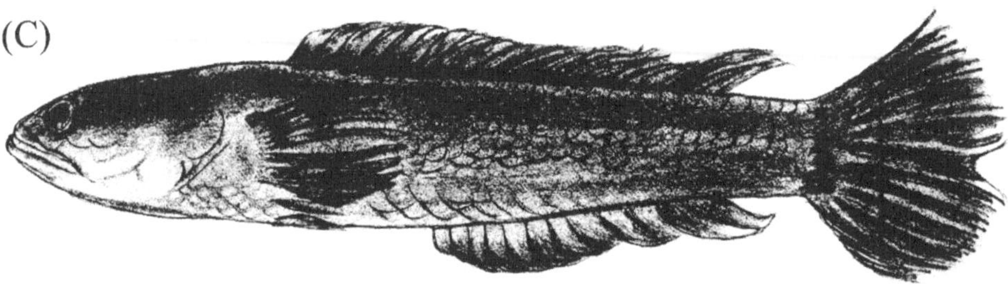

(D)

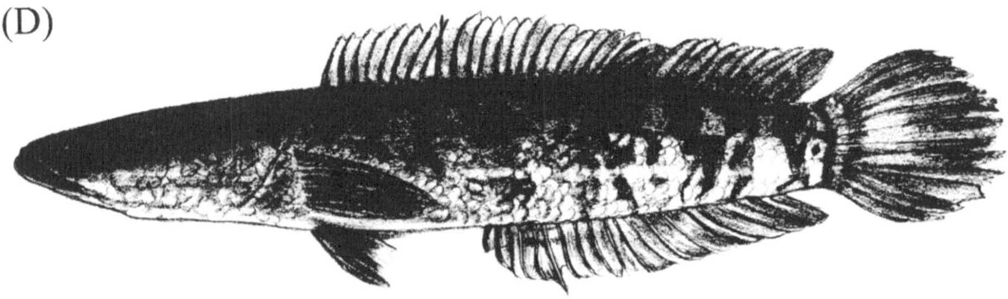

Plate 1a: (A) *Channa marulius*, (B) *Channa punctata*, (C) *Channa striata*, (D) *Channa gachua* collected form the river Gomti.

Order: PERCIFORMES

Family: Gobiidae (Gobies) Plate 2 and 2a

Species sampled	*Glossogobius giuris*
Collection of samples	Ganga River, site: Narora (28.1968° N, 78.3814° E)
Habitat and Distribution (FishBase)	Most inland freshwater bodies of India and also common in the coastal and estuarine waters from austral Africa and Madagascar to India and south of the China..
General Characters (Jayaram 1999)	Body varying from oblong to very elongate, eel like in some genera. Head generally with mucous canals and open pores. Five branchiostegal rays. Two dorsal fin united with each other or separated narrowly by a notch, or well separated, with soft flexible spines and rays. Pelvic fin united, usually forming an adhesive disc. Basal membrane present or absent. Anal fin short or long, or continuous with caudal fin. Anal papilla present. Lateral line absent; exposed pit-organs present. Scales ctenoid or cycloid.
Conservation status (IUCN)	Least concern
Otolith Morphometry	Otolith shape: approximately square, Thickness: moderate, Dorsal margin: slightly rounded, lobed. Ventral margin: flat, Posterior margin: emarginate. Rostrum: absent. Antirostrum: absent. Excisura: absent.

Family-Gobiidae

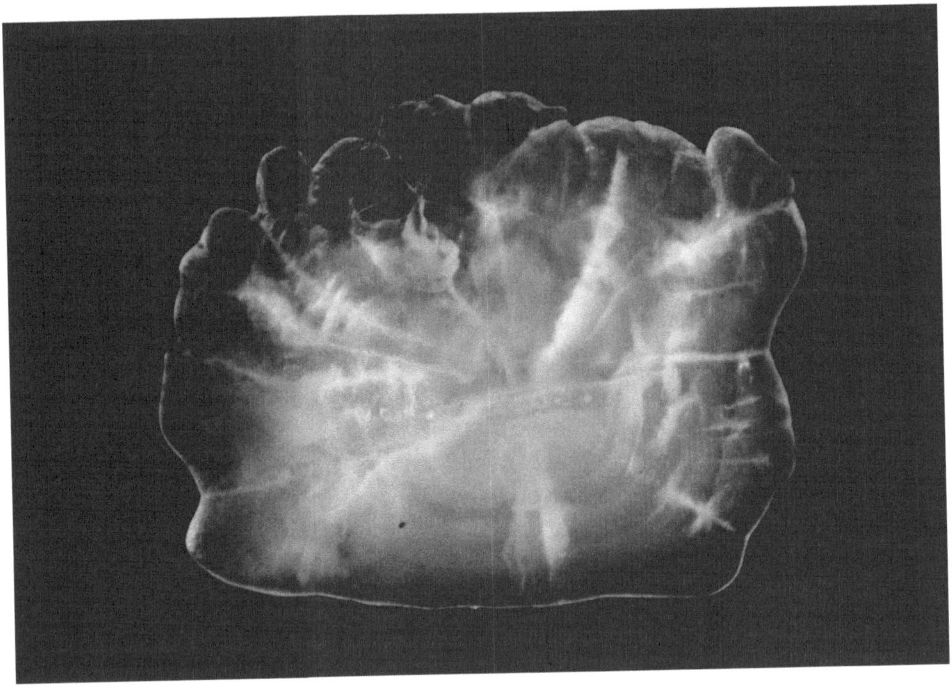

Plate 2: Otolith (sagitta) of *Glossogobius giuris* (OL=3.38 mm; OH=2.68 mm) collected form the river Ganga.

Family– Gobiidae

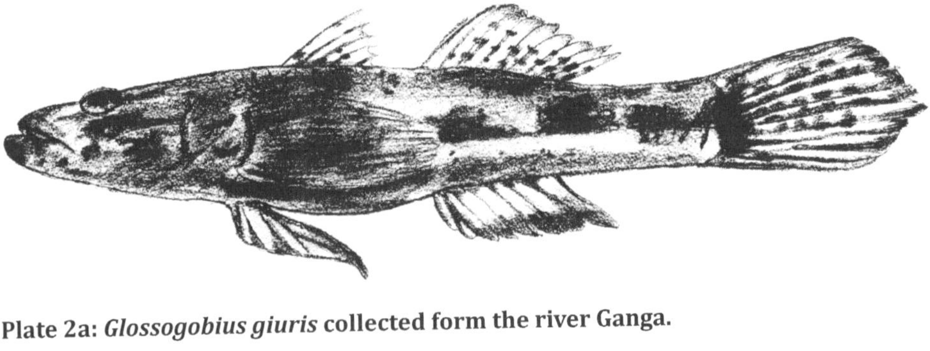

Plate 2a: *Glossogobius giuris* collected form the river Ganga.

Order: PERCIFORMES

Family: Nandidae (Leaf fishes) - Plate 3 and Plate 3a

Species sampled	*Nandus nandus*
Collection of samples	Yamuna River, site: Agra (27.1767° N, 78.0081° E)
Habitat and Distribution (FishBase)	Mainly distributed in the freshwaters of Asia: India, Pakistan to Thailand. However, other members of the family are reported from South America, West Africa and Southern Asia.
General Characters (Jayaram 1999)	Body oblong, elevated compressed with moderate sized ctenoid scales. Head covered with scales. Teeth on jaws and palate. Branchiostegal rays five or six. Operculum armed with a spine. A single dorsal fin with a spinous and a soft portion. Pelvic fins thoracic, without thoracic process. Anal fin with spines. Lateral line interrupted or absent. Air bladder present.
Conservation status (IUCN)	Least concern
Otolith morphology	Otolith shape: Oblong to fusiform, Thickness: moderately thin, Dorsal margin: gently rounded, slightly irregular and lobed, Ventral margin: slightly rounded, serrate in middle, Posterior margin: rounded and lobed, Rostrum: large and lobed, Antirostrum: small and pointed, Excisura: wide, notch shallow, angle wide.

Family-Nandidae

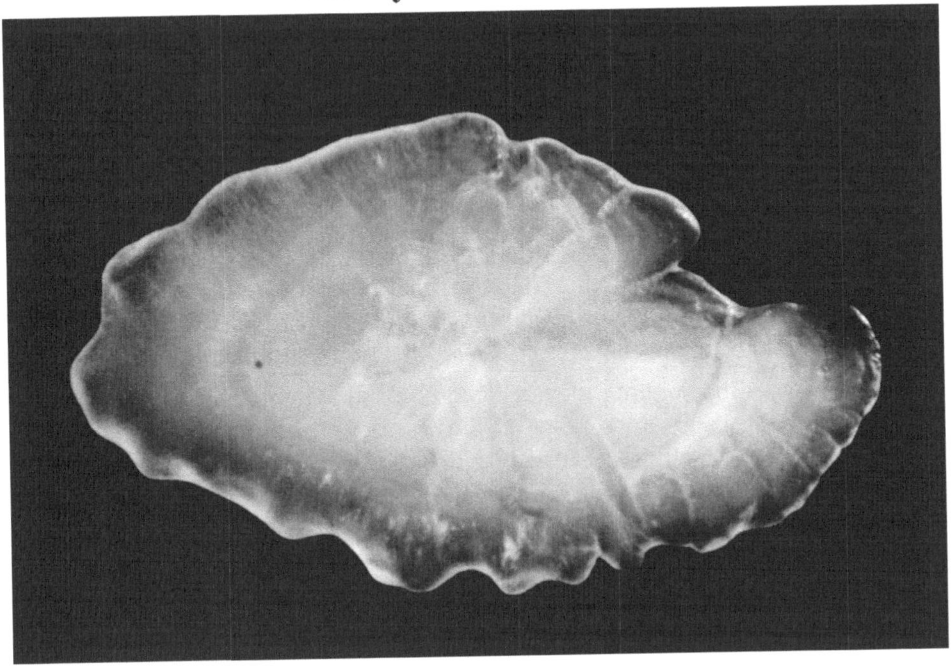

Plate 3: Otolith (sagitta) of *Nandus nandus* (OL=3.21 mm; OH=1.94 mm) collected form the river Yamuna.

Family– Nandidae

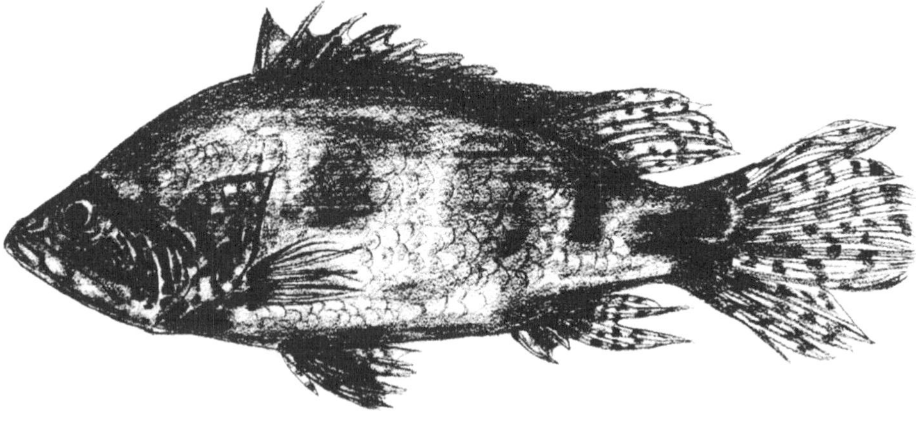

Plate 3a: *Nandus nandus* (TL= 15.3 cm) collected form the river Yamuna.

Order: PERCIFORMES

Family: - Ambassidae (Glassfishes) -Plate 4 and Plate 4a

Species sampled	*Parambassis ranga*
Collection of samples	Ganga River, site: Narora (28.1968° N, 78.3814° E)
Habita and Distribution (FishBase)	Freshwater and brackish water environments of Asia: India, Pakistan, Bangladesh, Myanmar, Thailand, Malaysia.
General Characters (Jayaram 1999)	Body short, elevated, oblong, compressed, slightly translucent. Moderate to small, deciduous cycloid scales on head and body. Branchiostegal rays six. Opercle without spine, ending in a membranous flap. Two dorsal fins, with a spinous and soft part. Anal fin with spines. A forwardly directed recumbent spine in front of dorsal fin present. Lateral line complete or interrupted.
Conservation status (IUCN)	Least concern
Otolith morphology	Otolith shape: slightly ovate, Thickness: moderately thick, Dorsal margin: well-rounded with a hump, Ventral margin: slightly flat with a small depression, Posterior margin: rounded and entire, Rostrum: broad, rounded and pointed. Antirostrum: small and slightly rounded, Excisura: very shallow, angle wide.

Family-Ambassidae

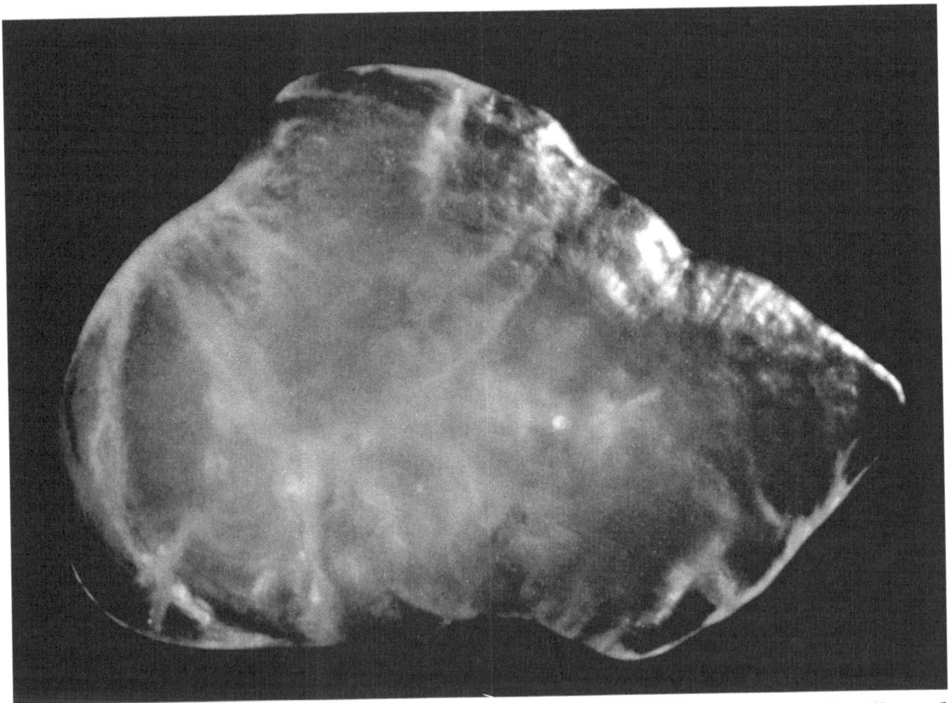

Plate 4: Otolith (sagitta) of *Parambassis ranga* (OL=0.85 mm; OH=0.61 mm) collected form the river Ganga.

Family– Ambassidae

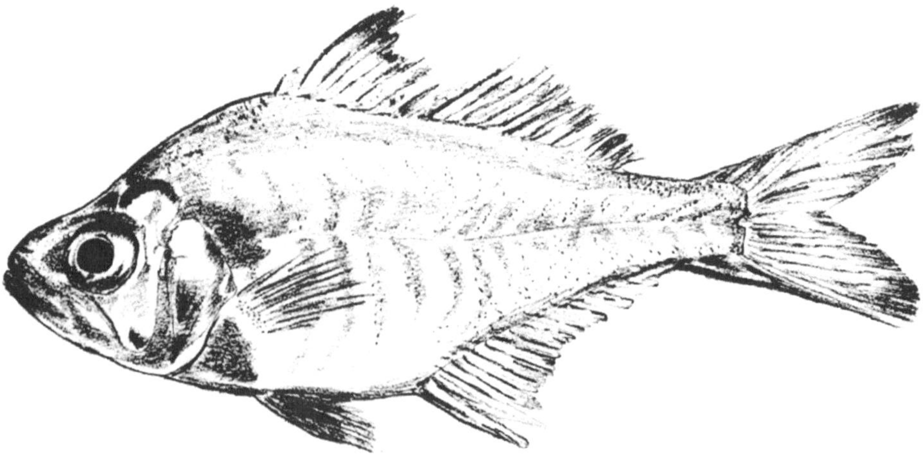

Plate 4a: *Parambassis ranga* collected form the river Ganga.

Order: SILURIFORMES

Family: Schilbeidae (Schilbid catfishes) – Plate 5 and Plate 5a

Species sampled	*Ailia coila, Clupisoma garua*
Collection of samples	Gomti River, site: Lucknow (26.8467° N, 80.9462° E)
Habitat and Distribution (FishBase)	Mainly found in the freshwater environments of India, Africa, Bangladesh, Malaya, Myanmar, Pakistan, Thailand.
General Characters (Jayaram 1999)	Medium to larger sized fishes with a compressed body. Teeth on pre-maxillaries, mandible. Barbels two, four, or eight, fairly developed. Gill opening very wide, extending up to lateral line, membranes free from each other and also from isthmus. Branchiostegal rays 5 to 12. Rayed dorsal fin when present short, with five to seven rays and a spine, may be absent also. Adipose fin generally present, may be absent. Pectoral fin with a strong spine, usually serrated. Anal fin very long, not confluent with caudal fin. Caudal fin forked. Lateral line generally complete, simple.
Conservation status	*Ailia colia* (Near Threatened, IUCN 2011), *Clupisoma garua* (Least Concern, IUCN 2010)
Otolith morphology	Otolith shape: ovate to rectangular, Thickness: moderate, Dorsal margin: slightly flat and irregular in the middle, Ventral margin: slightly rounded with one depression, Posterior margin: slightly emarginate, Rostrum: short, broad and rounded, Antirostrum: small, broad and rounded, Excisura: moderate, notch shallow, angle wide.

Family-Schilbeidae

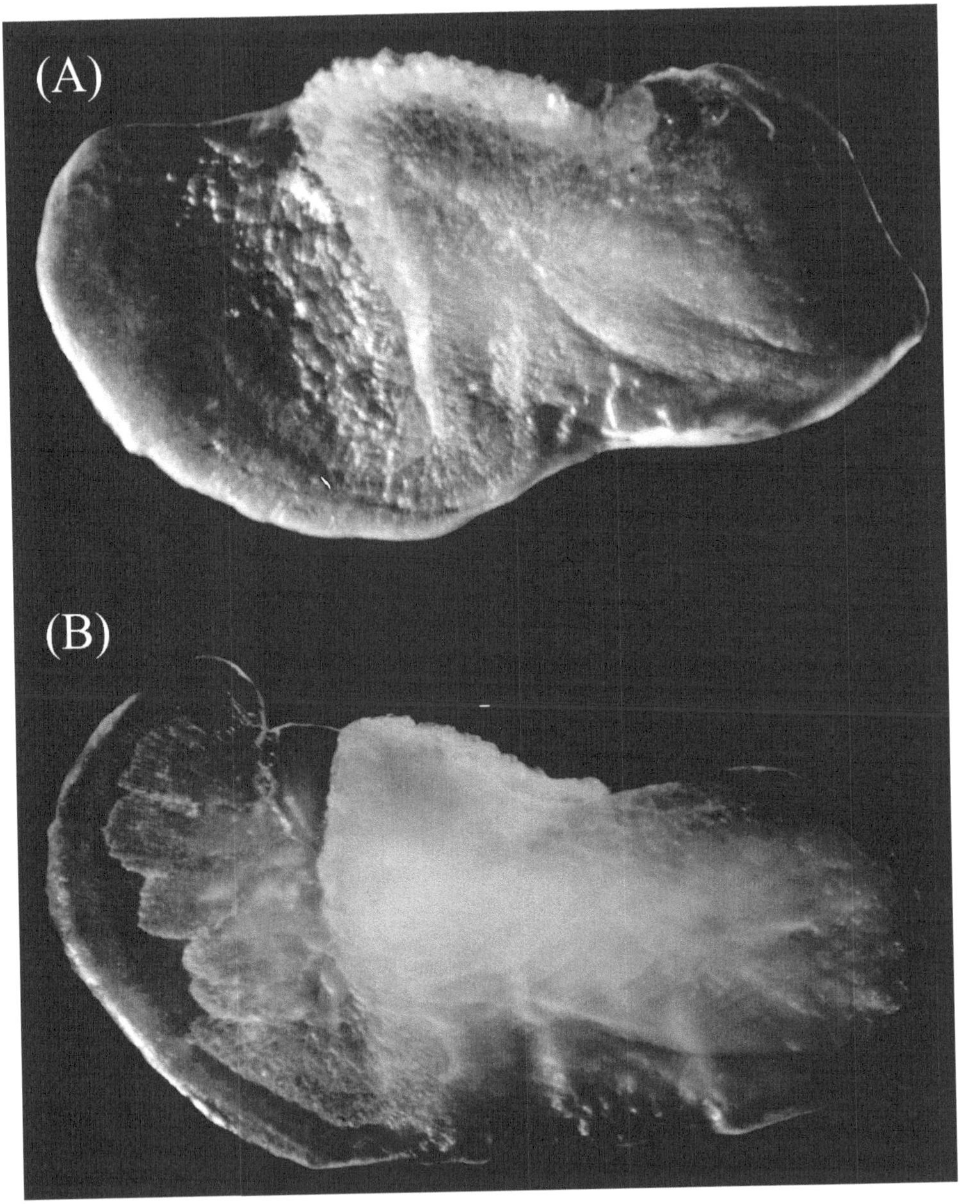

Plate 5: Otolith (lapillus) of (A) *Ailia coila* (OL=1.33 mm; OH=0.75 mm), (B) *Clupisoma garua* (OL=2.03 mm; OH=1.27 mm) collected form the river Gomti.

Family– Schilbeidae

(A)

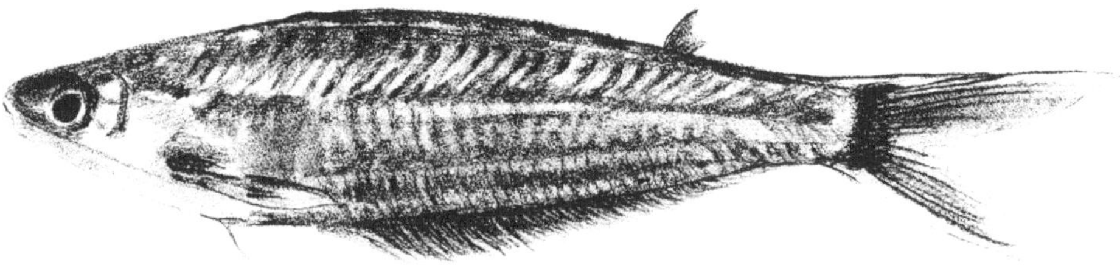

(B)

Plate 5a: (A) *Ailia coila*, (B) *Clupisoma garua* collected form the river Gomti.

Order: SILURIFORMES

Family: Clariidae (Air breathing catfishes) - Plate 6 and Plate 6a

Species sampled	*Clarias batrachus, Clarias gariepinus*
Collection of samples	Yamuna River, site: Agra (27.1767° N, 78.0081° E)
Habitat and Distribution (FishBase)	Freshwater environments of Asia and Africa.
General Characters (Jayaram 1999)	Large sized elongate fishes with a compressed body. Teeth on premaxillaries, mandibles and vomer. Barbels eight well developed. Gill opening wide extending to above base of pectoral fins. Branchiostegal rays 7 to 9. Rayed dorsal fin very long, separate or continuous with caudal fin without a spine. No adipose dorsal fin. Pectoral fin with a strong spine, may be serrated; in some genera fin vestigial. Anal fin long not confluent with caudal. Caudal fin rounded. Lateral line present, complete.
Conservation status	*Clarias batrachus* (Least concern, IUCN 2011), *Clarias gariepinus* (Least concern, IUCN 2016),
Otolith morphology	Otolith shape: Square, lobed and serrated. Thickness: moderately thick. Dorsal margin: slightly flat; sinuate or lobed. Ventral margin: Flattened and slightly crenate. Posterior margin: slightly rounded. Rostrum: Absent. Antirostrum: Absent. Excisura: Absent.

Family-Clariidae

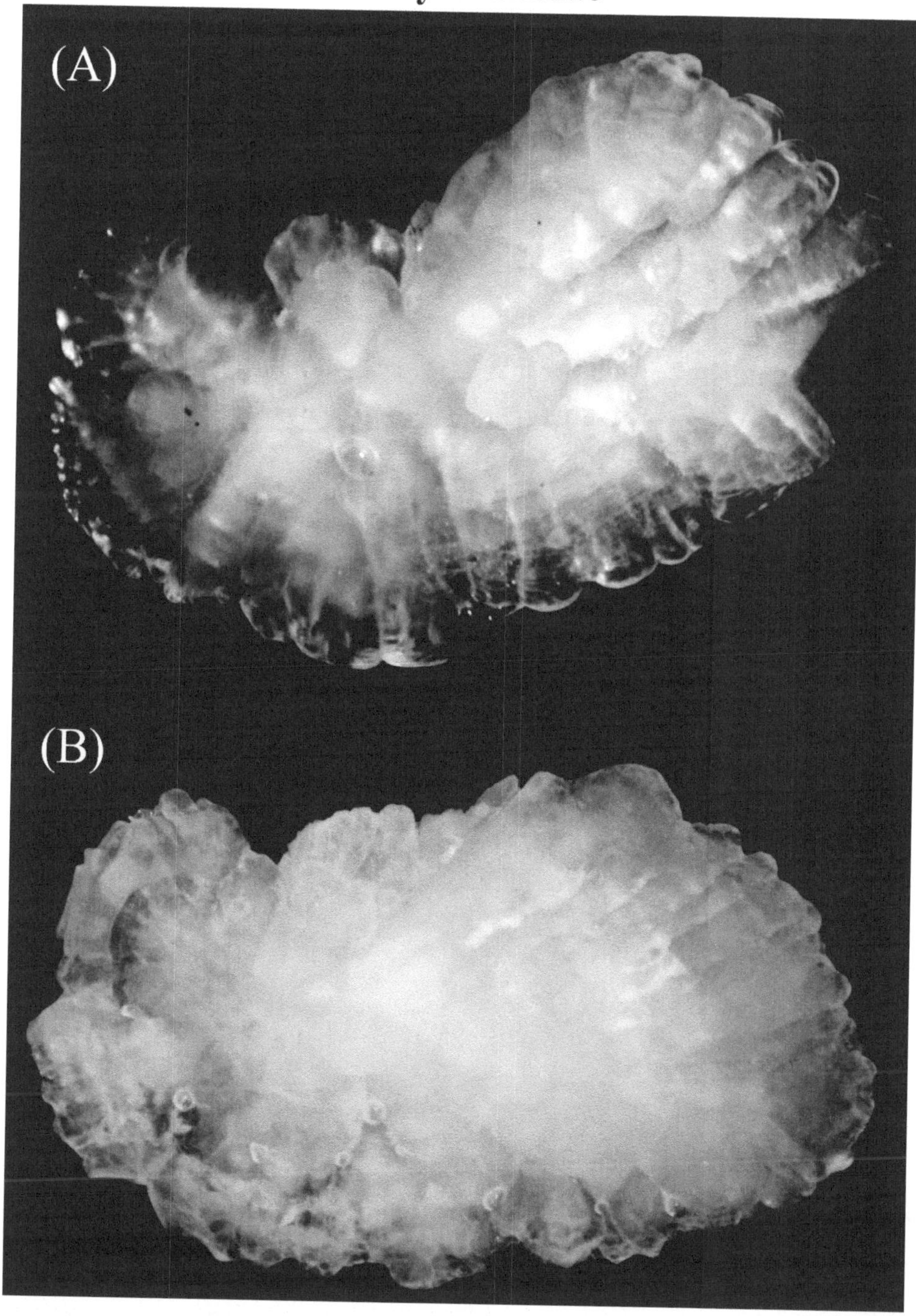

Plate 6: Otolith (lapillus) of (A) *Clarias batrachus* (OL=2.78 mm; OH=1.22 mm), (B) *Clarias gariepinus* (OL=4.37 mm; OH=3.12 mm), collected form the river Yamuna.

Family– Clariidae

(A)

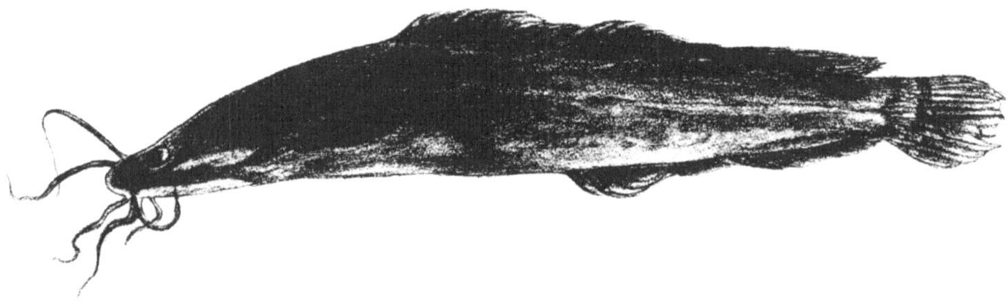

(B)

Plate 6a: (A) *Clarias batrachus*, (B) *Clarias gariepinus* collected form the river Yamuna.

Order: SILURIFORMES

Family: Sisoridae (Sisorid catfishes)- Plate 7 and Plate 7a

Species sampled	*Gagata cenia, Glyptothorax telchitta, Bagarius bagarius*
Habitat and Distribution (FishBase)	Freshwaters of Asia: India, Bangladesh, China, Malaya, Myanmar, Nepal, Pakistan, Thailand, Vietnam.
Collection of samples	Gomti River, site: Lucknow (26.8467° N, 80.9462° E)
General Characters (Jayaram 1999)	Generally small to medium sized fishes with head and anterior part of body depressed, and tail compressed or entire body compressed. An adhesive apparatus on thorax may be present or absent. Barbels eight, generally, well developed, in some genera thick, fleshy with broad bases. Rayed dorsal fin short, with or without a spine. Adipose dorsal fin smooth, generally free, but in some genera generally caudal, in some rudimentary. Pectoral fin with or without a spine. Anal fin short not confluent with caudal. Caudal fin deeply forked, emarginated, truncated or rounded. Lateral line present, complete.
Conservation status	*Gagata cenia, Glyptothorax telchitta,* (Least concern IUCN 2010), *Bagarius bagarius* (Near Threatened, IUCN 2010)
Otolith morphology	Otolith shape: slightly triangular, Thickness: moderately thick. Dorsal margin: flattened to irregular, Ventral margin: sharply rounded and slightly irregular, Posterior margin: V-shaped and irregular, Rostrum: small, broad and pointed, Antirostrum: small and sharply rounded, Excisura: Narrow, notch shallow, angle wide.

Family-Sisoridae

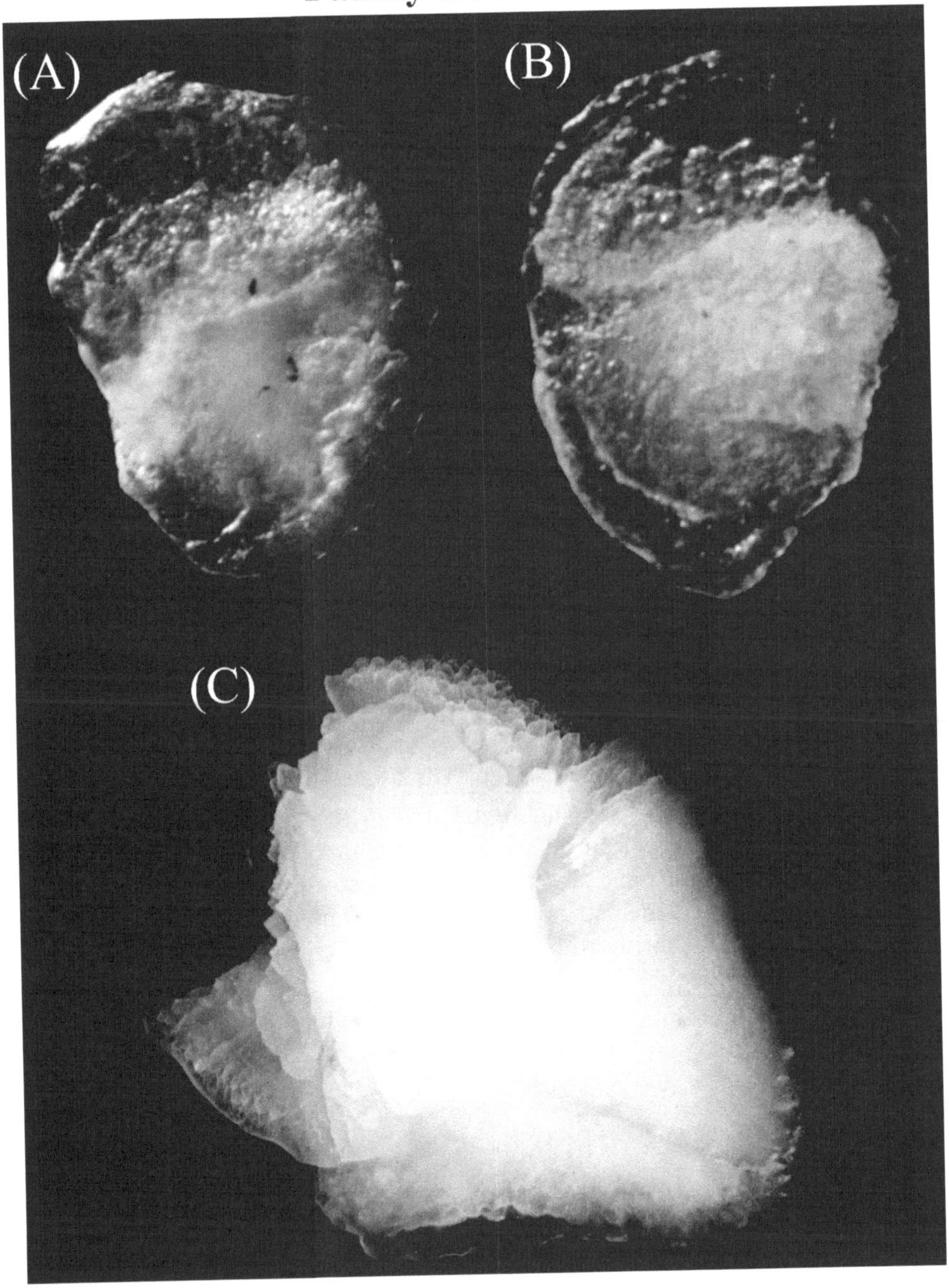

Plate 7: Otolith (lapillus) of (A) *Gagata cenia* (OL=0.50 mm; OH=0.43 mm), (B) *Glyptothorax telchitta* (OL=0.41 mm; OH=0.37 mm), (C) *Bagarius bagarius* (OL=6.37 mm; OH=4.81 mm) collected form the river Gomti.

Family– Sisoridae

(A)

(B)

(C)

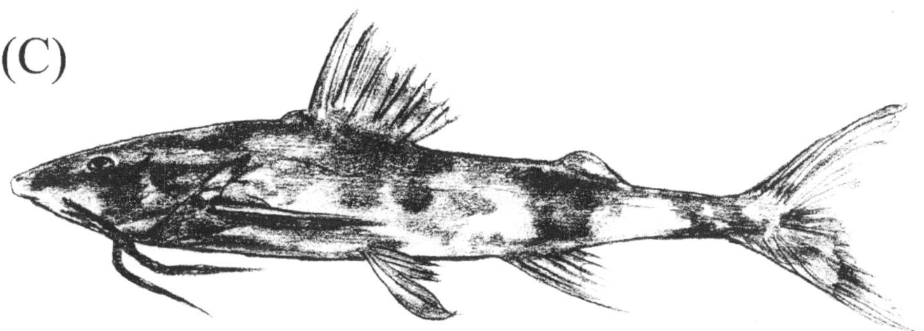

Plate 7a: (A) *Gagata cenia*, (B) *Glyptothorax telchitta*, (C) *Bagarius bagarius* collected form the river Gomti.

Order: SILURIFORMES

Family: Heteropneustidae (Stinging catfishes)- Plate 8 and Plate 8a

Species sampled	*Heteropneustus fossilis*
Collection of samples	Yamuna River, site: Agra (27.1767° N, 78.0081° E)
Habitat and Distribution (FishBase)	Freshwaters of India, Bangladesh, Myanmar, Pakistan, Sri Lanka, Thailand, Vietnam; and rarely in the brackish waters.
General Characters (Jayaram 1999)	Moderate sized elongate fishes with a compressed body. Teeth on premaxillaries, mandible and vomer. Barbels eight, well developed. Gill opening wide, extending to above and base of pectoral fins. A long air-sac extends posteriorly from gill-chamber through muscles of back to tail. Brannchiostegal rays seven. Rayed dorsal fin short and without any spine. Pectoral fin inserted horizontally. Adipose dorsal fin absent. Caudal fin almost rounded. Lateral line present, complete.
Conservation status (IUCN)	Least concern
Otolith morphology	Otolith shape: Rectangular and irregular, Thickness: moderately thick. Dorsal margin: flat and sinuate, Ventral margin: slightly flattened and irregular, Posterior margin: sharply rounded and slightly irregular. Rostrum: absent. Antirostrum: absent, Excisura: absent.

Family-Heteropneustidae

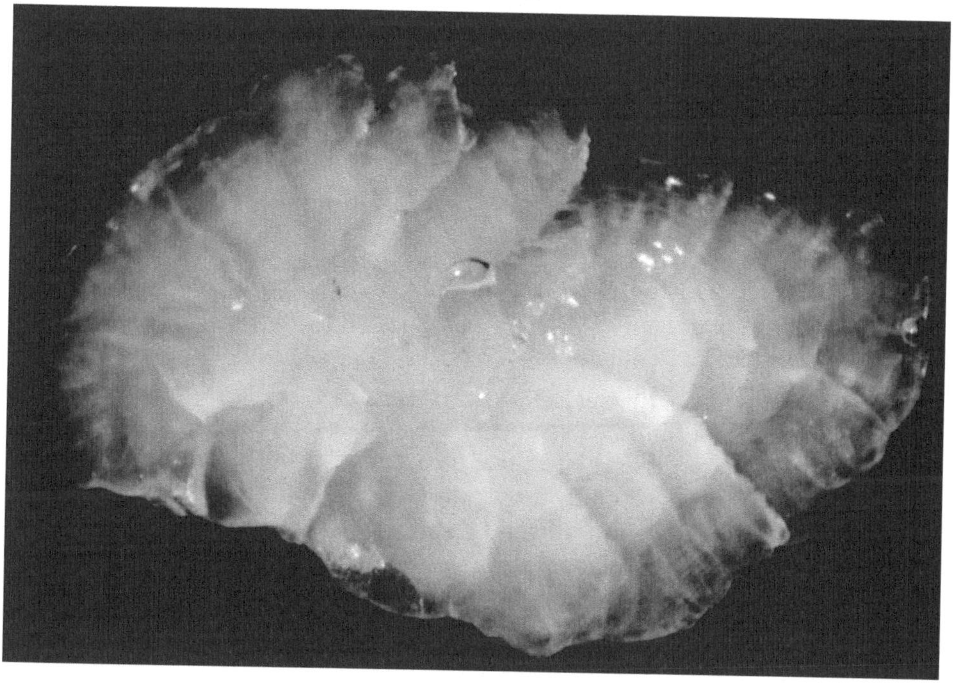

Plate 8: Otolith (lapillus) of (A) *Heteropneustus fossilis* (OL=2.67 mm; OH=1.59 mm) collected form the river Yamuna.

Family– Heteropneustidae

Plate 8a: (A) *Heteropneustus fossilis* collected form the river Yamuna.

Order: SILURIFORMES

Family: Bagridae (Bagrid catfish) Plate 9 and Plate 9a

Species sampled	*Sperata aor, Mystus cavasius, Mystus tengra, Sperata seenghala, Mystus vittatus, Rita rita*
Collection of samples	Gomti River, site: Lucknow (26.8467° N, 80.9462° E)
Habitat and Distribution (FishBase)	Freshwaters of Africa and Asia: ranging from Bangladesh, Cambodia, China, East Indies, Formosa, India, Japan, Korea, Laos, Malaya, Manchuria, Pakistan, Sri Lanka, Thailand, Vietnam and West Asia.
General Characters (Jayaram 1999)	Generally larger sized, more or less elongate fishes, with a compressed body. Teeth on premaxillaries, mandible and vomer. Nostrils widely separated, above angle of mouth, anterior tubular on tip of snout, posterior nearer eye than tip of snout and with the nasal barbels. Barbels six or eight pairs generally well developed. Gill opening wide, extending to above base of pectoral fins, membranes free from each other and also from the isthmus. Adipose dorsal fin smooth. Pectoral fin with strong spine. Caudal fin forked.
Conservation status	*Sperata aor* (Least concern IUCN 2011), *Mystus cavasius, Mystus tengra, Sperata seenghala,Mystus vittatus, Rita rita* (Least concern IUCN 2010)
Otolith morphology	Otolith shape: oval to irregular oblong, Thickness: moderately thin, Dorsal margin: grooved, Ventral margin: slightly rounded and slightly irregular, Posterior margin: rounded and irregular to lobed, Rostrum: absent, Antirostrum: absent, Excisura: absent.

Family-Bagridae

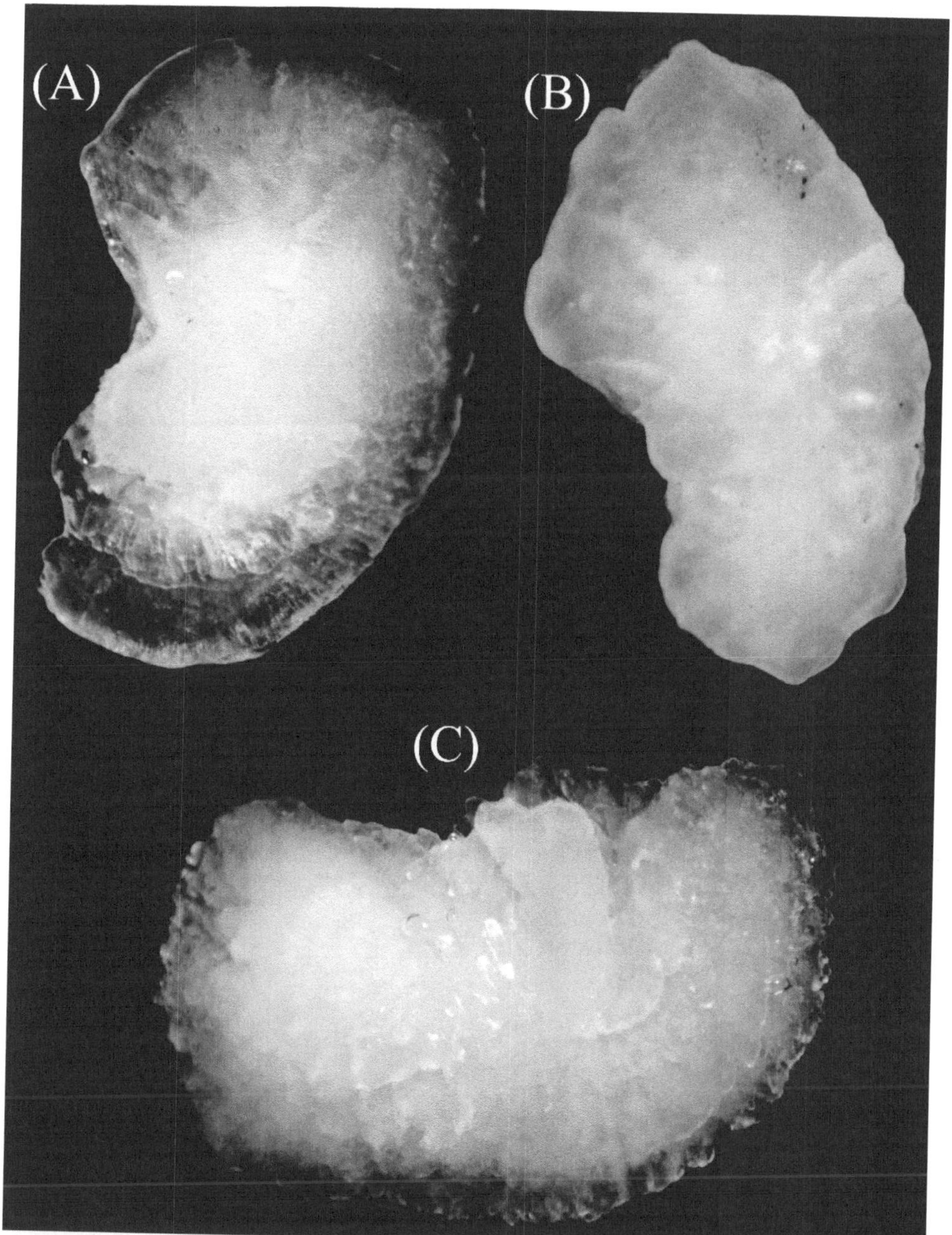

Plate 9: Otolith (lapillus) of (A) *Sperata aor* (OL=3.91 mm; OH=2.65 mm), (B) *Mystus cavasius* (OL=2.14 mm; OH=1.27 mm), (C) *Mystus tengra* (OL=2.31 mm; OH=1.22 mm) collected form the river Gomti.

Family-Bagridae

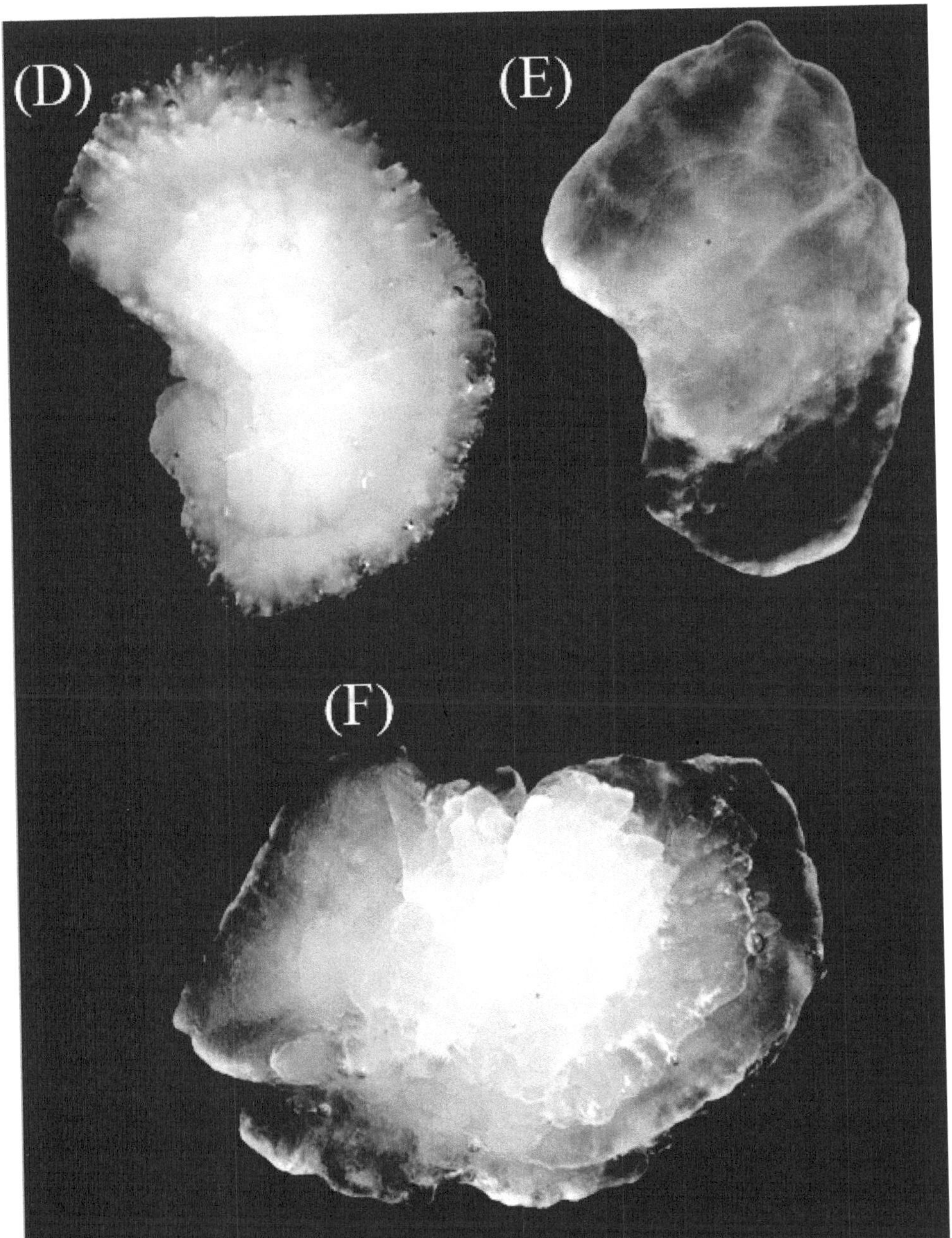

Plate 9: Otolith (lapillus) of (D) *Sperata seenghala* (OL=5.63 mm; OH=3.45 mm), (E) *Mystus vittatus* (OL=1.2 mm; OH=0.6 mm), (F) *Rita rita* (OL=3.08 mm; OH=1.93 mm) collected form the river Gomti.

Family– Bagridae

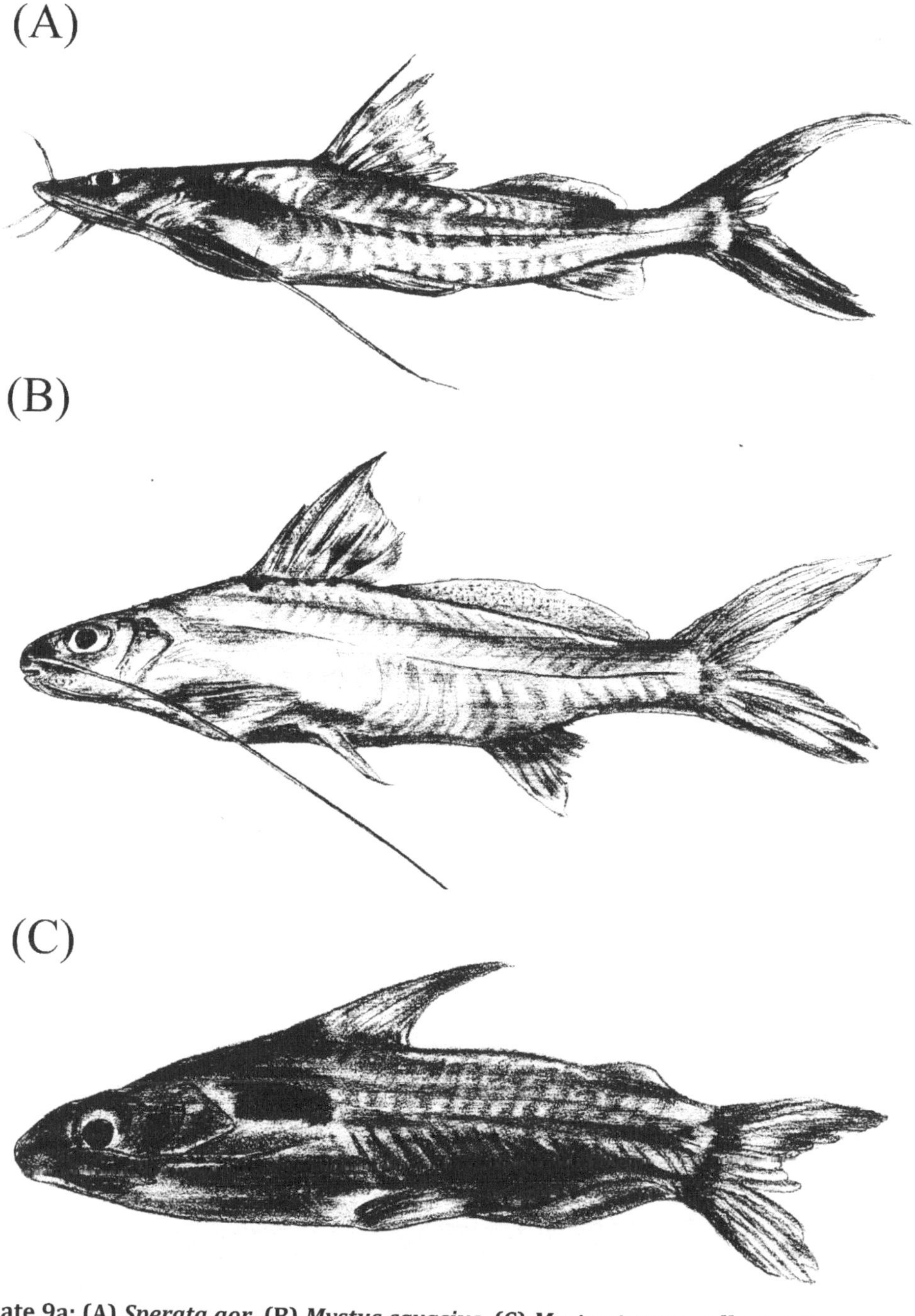

Plate 9a: (A) *Sperata aor*, (B) *Mystus cavasius*, (C) *Mystus tengra* collected form the river Gomti.

Family– Bagridae

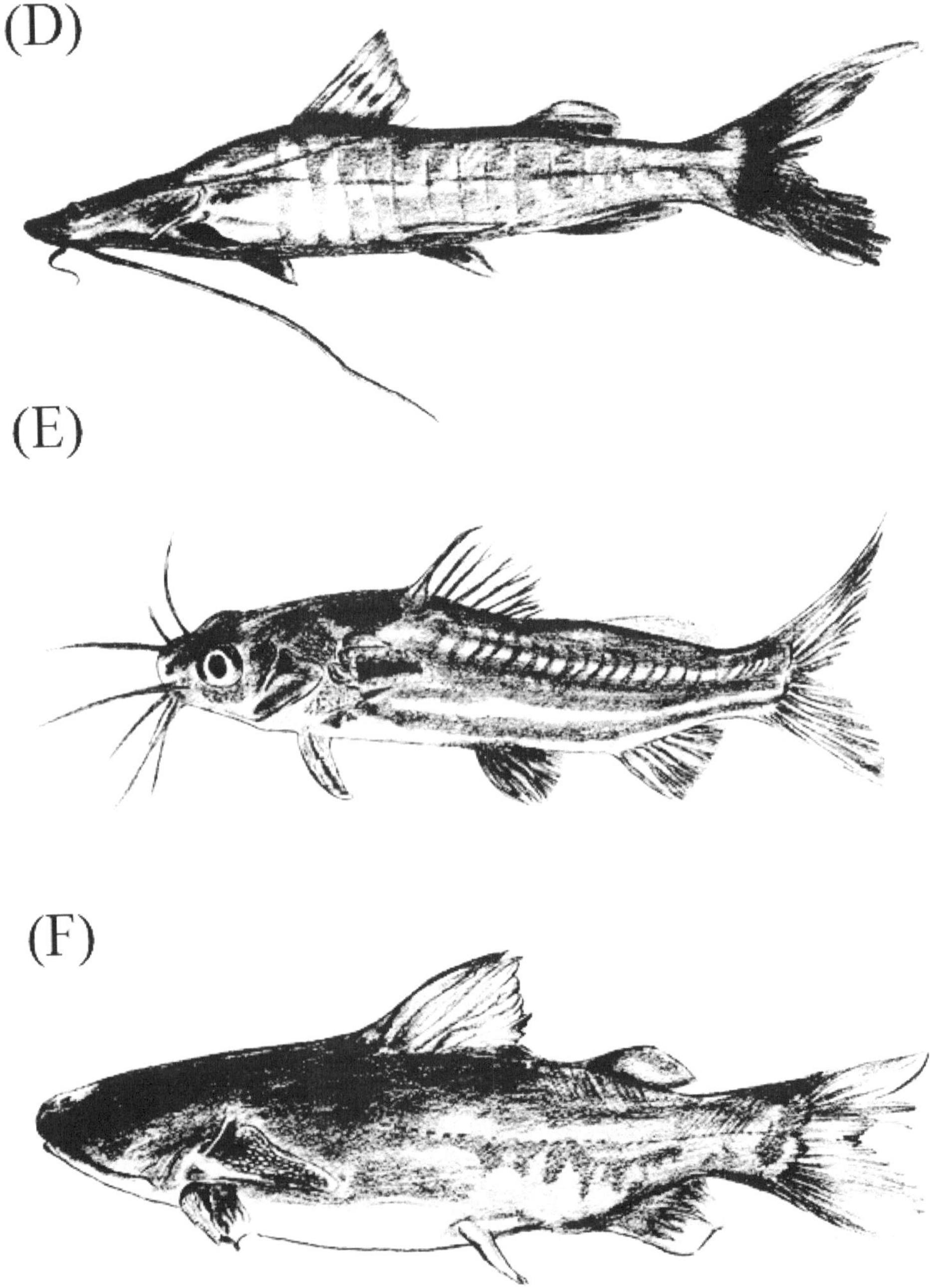

Plate 9a: (D) *Sperata seenghala*, (E) *Mystus vittatus*, (F) *Rita rita* collected form the river Gomti.

Order: SILURIFORMES

Family: Siluridae (Sheat fishes) Plate 10 and Plate 10a

Species sampled	*Wallago attu* and *Ompok pabda*
Collection of samples	Ganga River, site: Narora (28.1968° N, 78.3814° E)
Habitat and Distribution (FishBase 2018)	Freshwaters of India, Bangladesh, Indonesia, Malaya, Myanmar, Pakistan, Sri Lanka, Thailand, Vietnam.
General Characters (Jayaram 1999)	Larger sized elongate fishes with a compressed body. Depressible teeth on premaxillaries, mandible and vomer. Nostrils separated from each other by a short distance. Barbels four or six generally well developed. Nasal barbels invariably absent. Gill opening very wide. Brachiostegal rays 8 to 21. Rayed dorsal fin mostly one short with four or five rays and without a spine. Adipose dorsal fin absent. Caudal fin rounded to weakly emarginated, with bluntly rounded lobes or forked.
Conservation status (IUCN 2010)	Near Threatened
Otolith morphology	Otolith shape: rectangular and irregular, Thickness: moderately thick, Dorsal margin: slightly flat and lobed. Ventral margin: slightly rounded and irregular, Posterior margin: rounded to sharply rounded and irregular, Rostrum: sharply rounded and slightly irregular, Antirostrum: small, Excisura: moderate, notch very shallow, angle acute.

Family-Siluridae

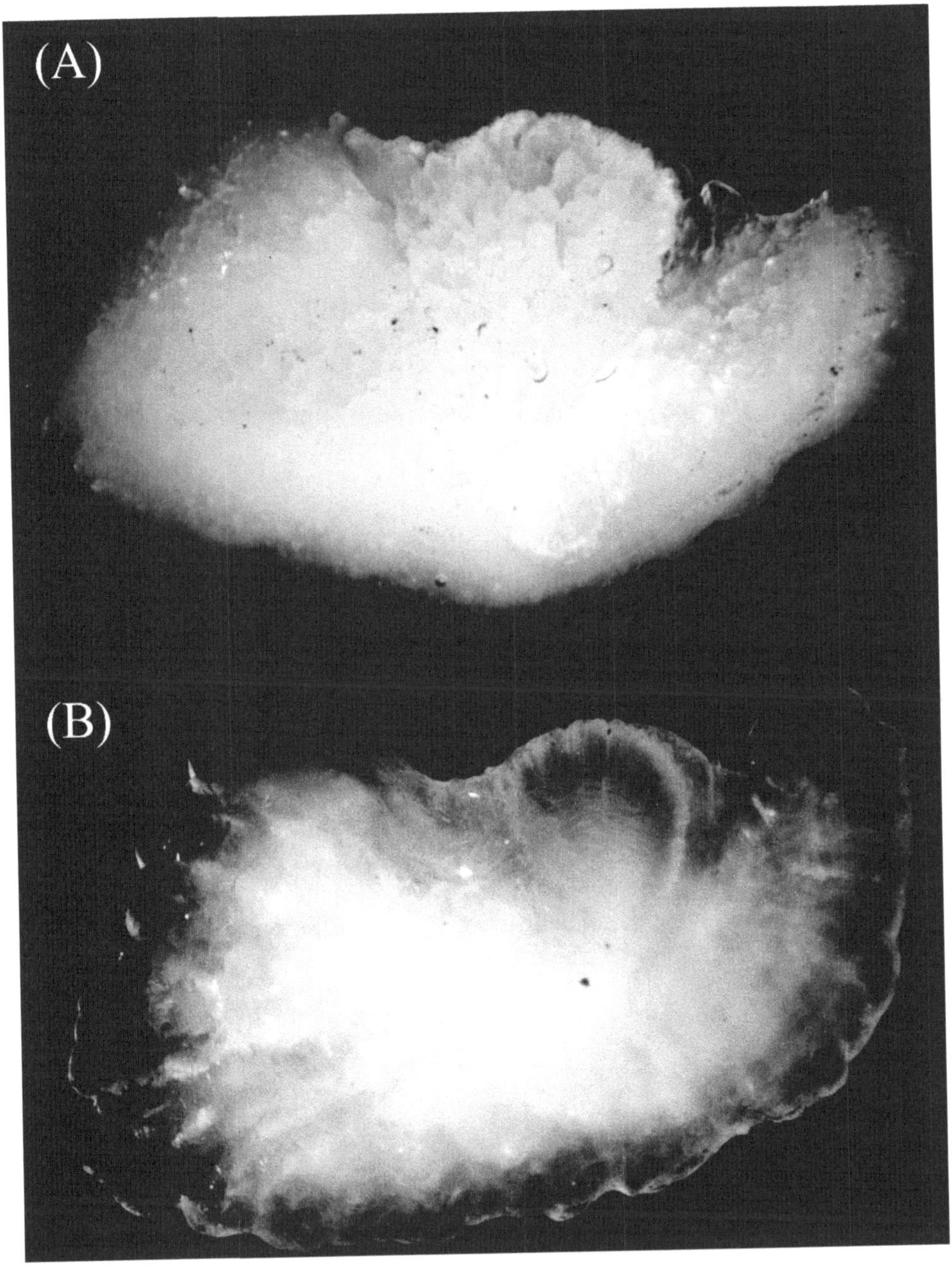

Plate 10: Otolith (lapillus) of (A) *Wallago attu* (OL=5.27 mm; OH=2.94 mm) and (B) *Ompok pabda* (OL=2.46 mm; OH=1.27) collected form the river Ganga.

Family– Siluridae

(A)

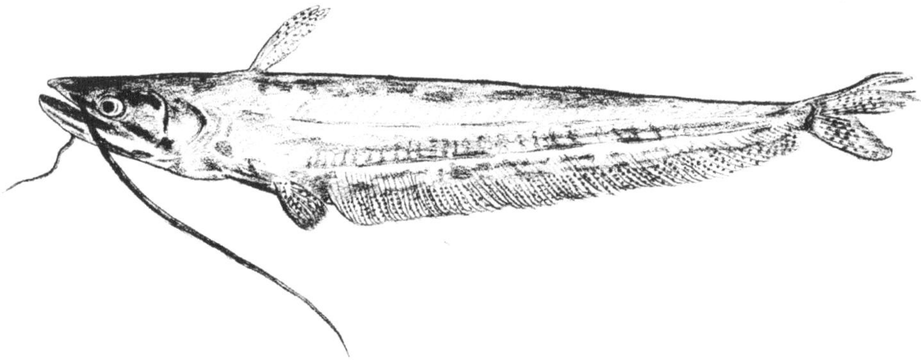

(B)

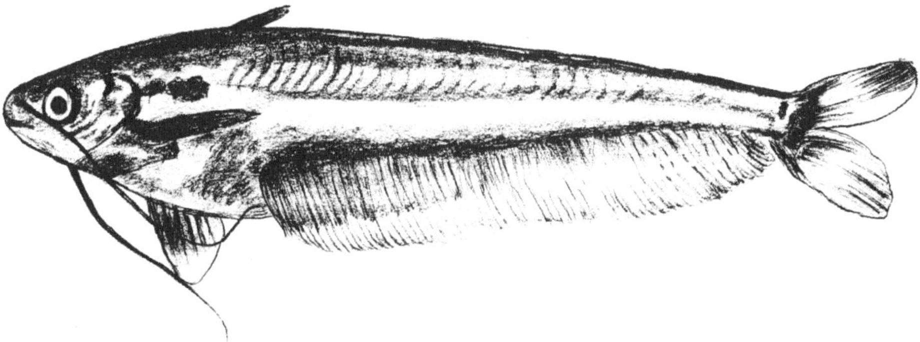

Plate 10a: (A) *Wallago attu* and (B) *Ompok pabda* collected form the river Ganga.

Order: CYPRINIFORMES

Family: Cobitidae - Plate 11 and Plate 11a

Species sampled	*Botia lohachata, Lepidocephalus thermalis*
Collection of samples	Ganga River, site: Narora (28.1968° N, 78.3814° E)
Habitat and Distribution (FishBase)	Asia: India, Pakistan, Nepal and Bangladesh and Sri Lanka
General Characters (Jayaram 1999)	Body short, low, more or less cylindrical, posteriorly compressed but never depressed. Mouth subterminal to inferior. Lips thick, fleshy and papillated. Three pairs of barbels. An erectile spine before or below eye present. Vertical fins without any spines. Pelvic fins may be absent. Lateral line present.
Conservation status	*Botia lohachata* (Not evaluated), *Lepidocephalus thermalis* (Least concern IUCN 2011)
Otolith morphology	Otolith shape: Approximately circular, Thickness: very thick and bulbous, Dorsal margin: round and entire. Ventral margin: slightly round and irregular in the middle Rostrum: absent, Antirostrum: absent, Excisura: absent.

Family-Cobitidae

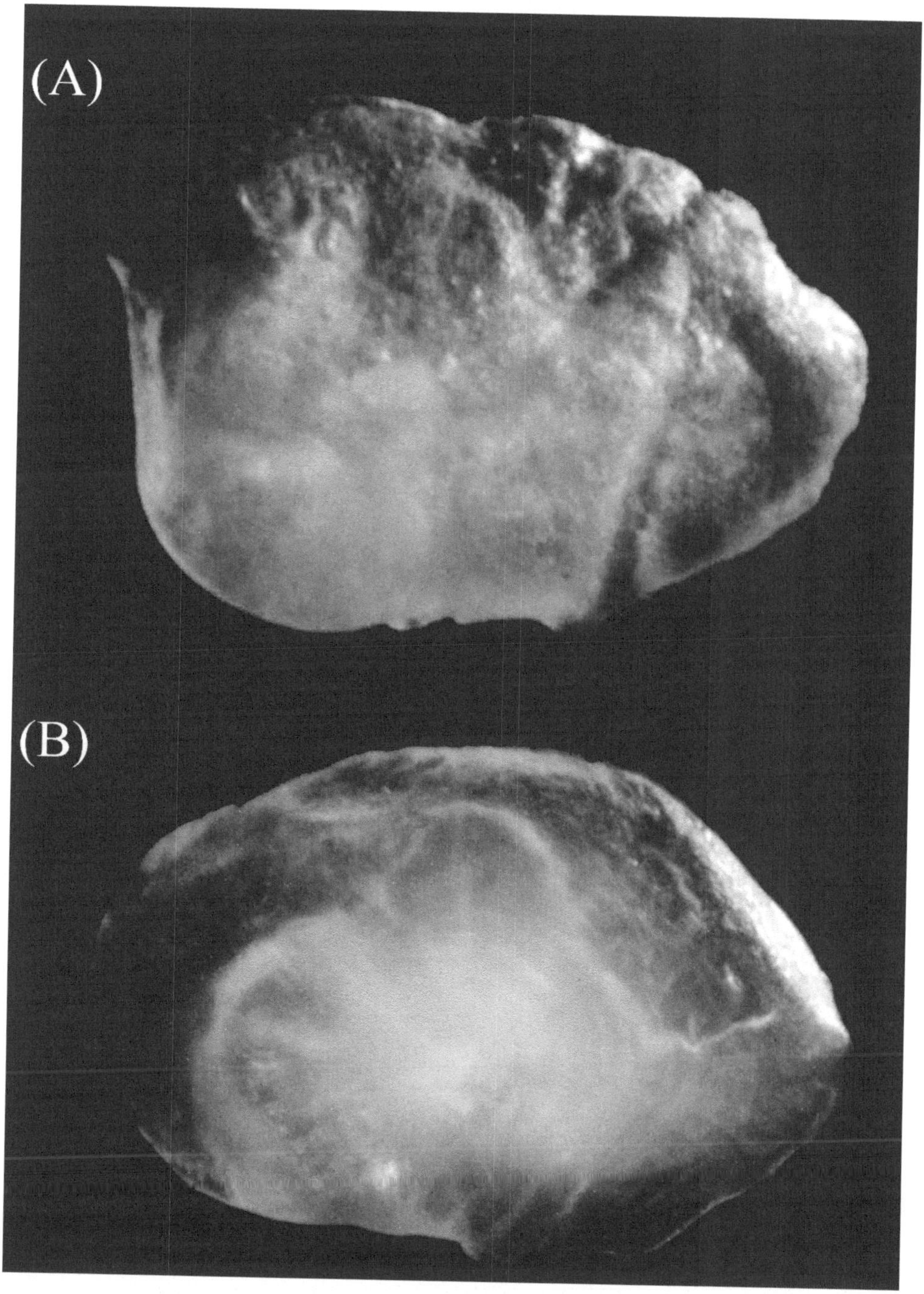

Plate 11: Otolith (asteriscus) of (A) *Botia lohachata* (OL=0.32 mm; OH=0.21 mm), (B) *Lepidocephalus thermalis* (OL=0.83 mm; OH=0.59 mm), collected form the river Ganga.

Family– Cobitidae

(A)

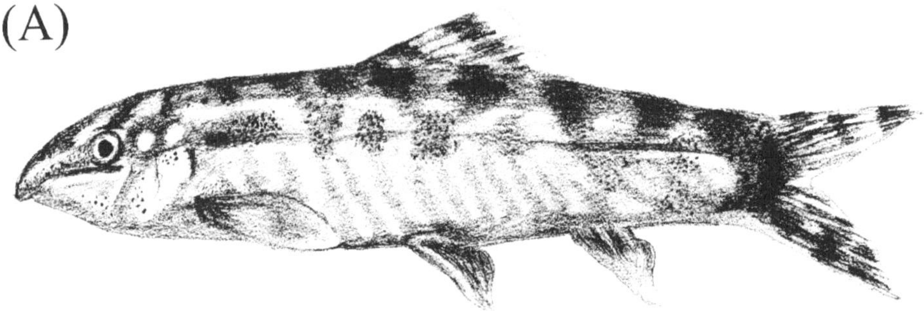

(B)

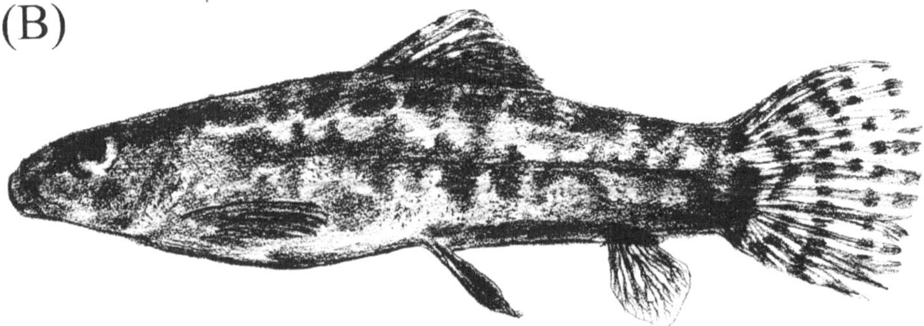

Plate 11a: (A) *Botia lohachata* (TL= 6.3 cm), (B) *Lepidocephalus thermalis* (TL= 6.8 cm), collected form the river Ganga.

Order: SYNBRANCHIFORMES

Family: Mastacembelidae- Plate 12 and Plate 12a

Species sampled	*Mastacembelus armatus, Macrognathus pancalus, Macrognathus aculeatus*
Collection of samples	Ganga River, site: Narora (28.1968° N, 78.3814° E)
Habitat and Distribution (FishBase 2018)	Freshwaters of Bangladesh, Cambodia, China, India, Myanmar, Nepal, Pakistan, Sri Lanka, Thailand, Vietnam
General Characters (Jayaram 1999)	Body eel-like, compressed and elongate, covered with minute scales. Mouth non-protractile. Snout elongate and supported by a cartilaginous rod and ending in a sensitive tip flanked by tubular nostrils anteriorly placed on the snout. Branchiostegals five or six. Dorsal fin long with 7 to 40 detached depressible dorsal. Pectoral fin with 17 to 27 rays. Caudal fin short, either confluent with dorsal and anal fins or narrowly separated; homocercal.
	Mastacembelus armatus, Macrognathus pancalus (Least Concern, IUCN 2010), *Macrognathus aculeatus* (*Not evaluated*)
Otolith morphology	Otolith shape: oblong, Thickness: moderately thin, Dorsal margin: slightly rounded and slightly entire, Ventral margin: slightly rounded and slightly irregular, Posterior margin: sharply rounded and entire, Rostrum: small and lobed, Antirostrum: small and rounded. Excisura: moderate, notch very shallow, angle wide.

Family-Mastacembelidae

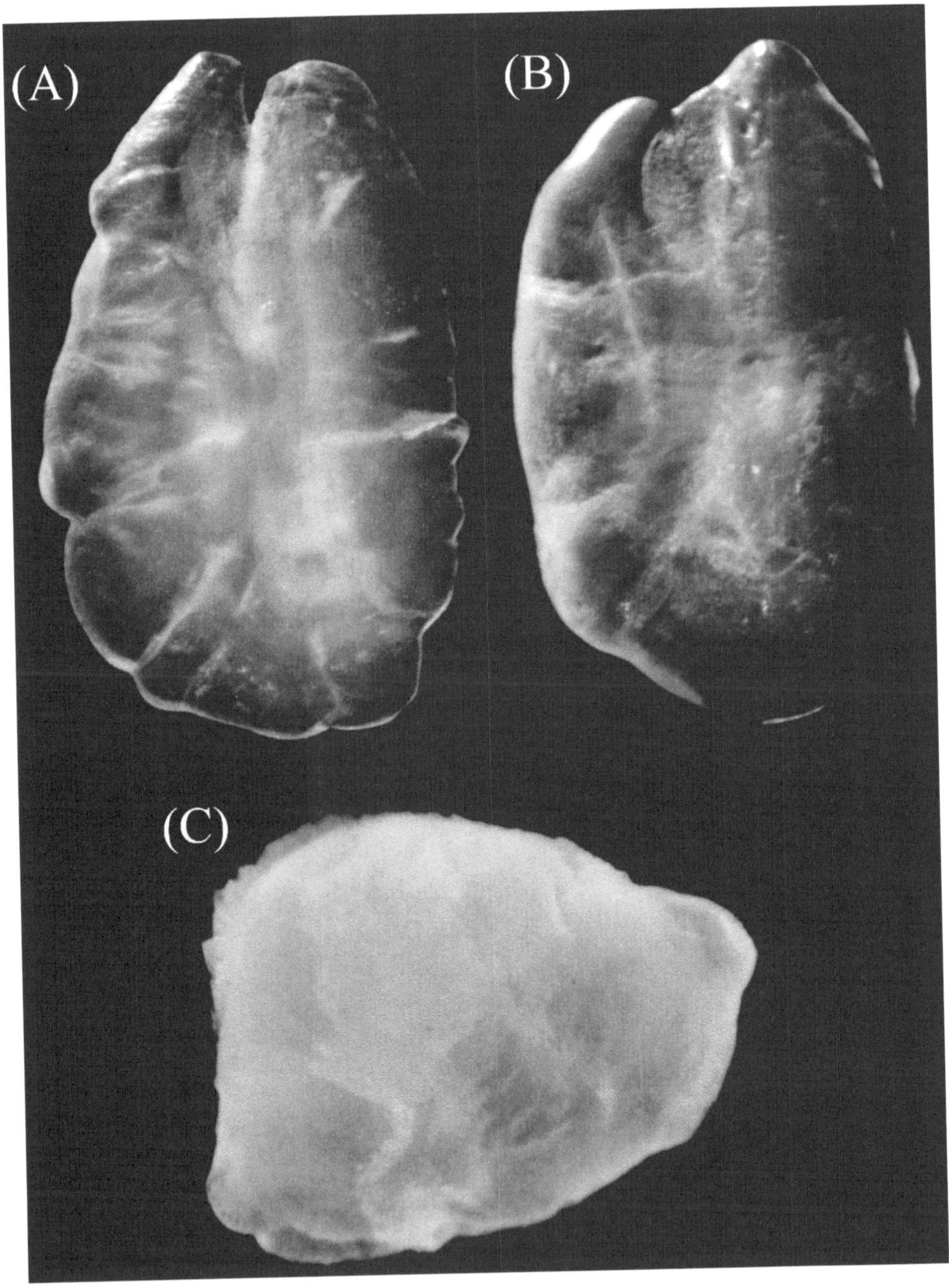

Plate 12: Otolith (sagitta) of (A) *Mastacembelus armatus* (OL=1.66 mm; OH=0.79 mm), (B) *Macrognathus pancalus* (OL=1.03 mm; OH=0.65 mm), (C) *Macrognathus aculeatus* (OL=0.48 mm; OH=0.39 mm) collected form the river Ganga.

Family– Mastacembelidae

(A)

(B)

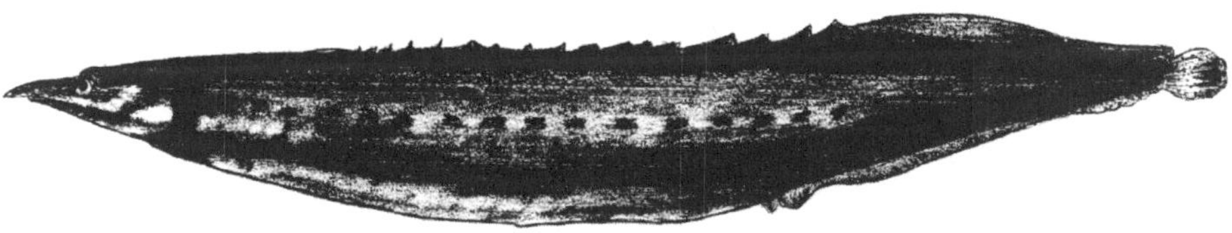

(C)

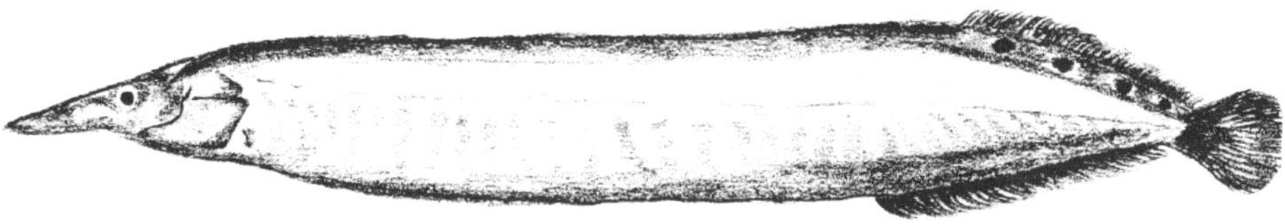

Plate 12a: (A) *Mastacembelus armatus*, (B) *Macrognathus pancalus*, (C) *Macrognathus aculeatus* collected form the river Ganga.

Order: BELONIFORMES

Family: Belonidae (Freshwaters Gars) - Plate 13 and Plate 13a

Species sampled	*Xenentodon cancila*
Collection of samples	Gomti River, site: Lucknow (26.8467° N, 80.9462° E)
Habitat and Distribution (FishBase)	Freshwaters of Asia: India eastward to the Mekong and Sri Lanka.
General Characters (Jayaram 1999)	Body elongate, slender, with small scales. Head with scales. Both jaws elongated as a beak armed with sharp teeth to their tip. Teeth on jaws; Branchiostegal rays 12 to 15. Gill opening wide. Dorsal fin placed far posterior of body with Identifying keys spines; Dorsal fin rays varying in elevation, sometimes forming a lobe to the fin. Pectoral fin shorts with 10 to 12 rays. Pelvic fin abdominal with six rays. Lateral line on free portion of tail with or without a keel. Scales cycloid, deciduous.
Conservation status (IUCN)	Least concern
Otolith morphology	Otolith shape: nearly oblong, Thickness: moderately thin, Dorsal margin: sharply rounded and irregular; Ventral margin: slightly flattened, irregular or crenate, Posterior margin: rounded, and crenate, Rostrum: small, broad and lobed. Antirostrum: small and sharply rounded, Excisura: narrow, notch shallow and angle acute.

Family-Belonidae

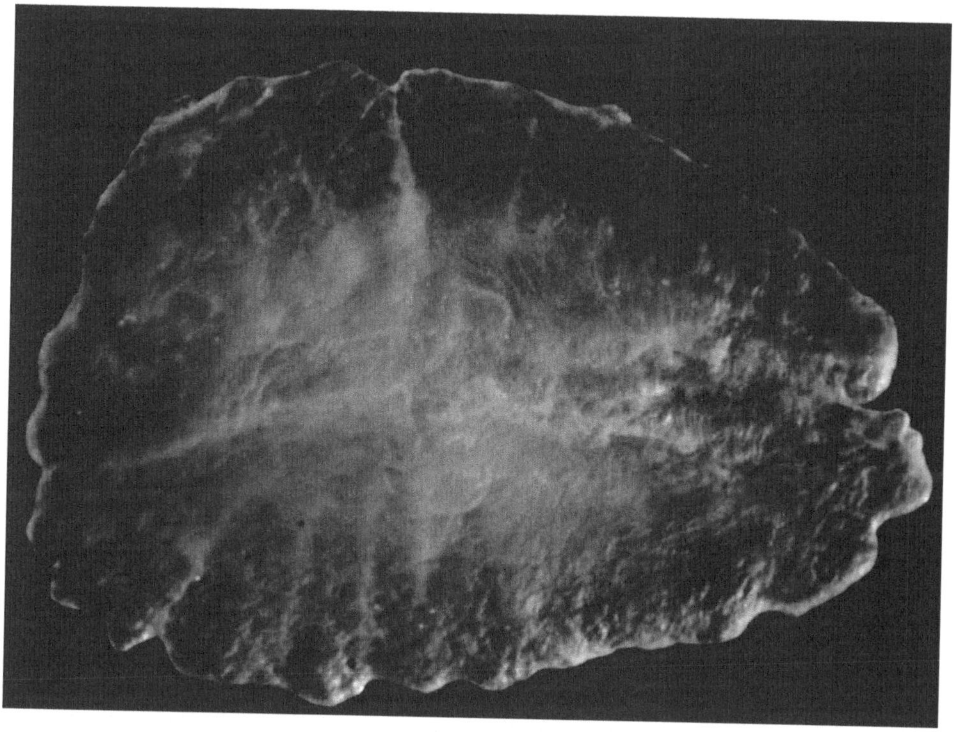

Plate 13: Otolith (sagitta) of *Xenentodon cancila* (OL=1.93 mm; OH=1.12 mm) collected form the river Gomti.

Family– Belonidae

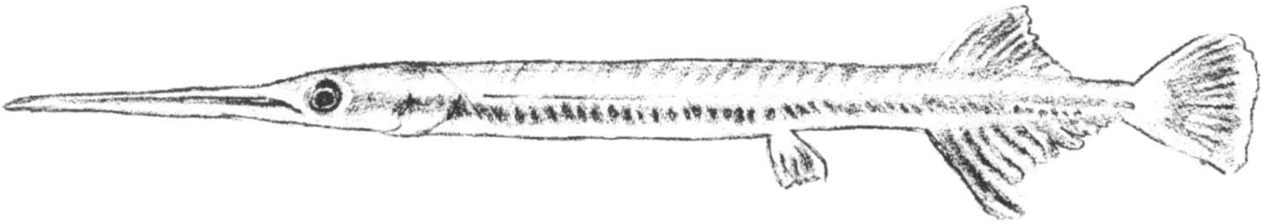

Plate 13a: *Xenentodon cancila* collected form the river Gomti.

Order: CLUPEIFORMES

Family: Clupidae - Plate 14 and Plate 14a

Species sampled	*Gadusia chapra*
Collection of samples	Ganga River, site: Narora (28.1968° N, 78.3814° E)
Habitat and Distribution (FishBase)	Asia: rivers of India and Bangladesh affluent to the Bay of Bengal (chiefly the Ganges and Brahmaputra system and Mahanadi river of Orissa). Reported from Nepal and Pakistan. However, members of the family have also been reported from the Freshwater, Brackish water and Marine environments of Tropical Africa.
General Characters (Jayaram 1999)	Maxillaries composed of three pieces not ossified together. Abdomen with scutes. Head without scales. Barbels absent. Gill membranes free. Opercular pieces four. A single dorsal fin with a few or a moderate number of weak rays. Scales present on body. Lateral line mostly absent. Pyloric appendages when present numerous.
Conservation status (IUCN)	Least concern
Otolith morphology	Otolith shape: Ovate, Thickness: moderately thick, Dorsal margin: slightly rounded and becoming flat in the middle and entire. Ventral margin: approximately flat and dentate, Posterior margin: slightly rounded and slightly irregular, Rostrum: moderate and sharply rounded becoming elongate and pointed, Antirostrum: large and becoming elongate and pointed, Excisura: wide and notch shallow becoming moderate, angle wide becoming acute.

Family-Clupidae

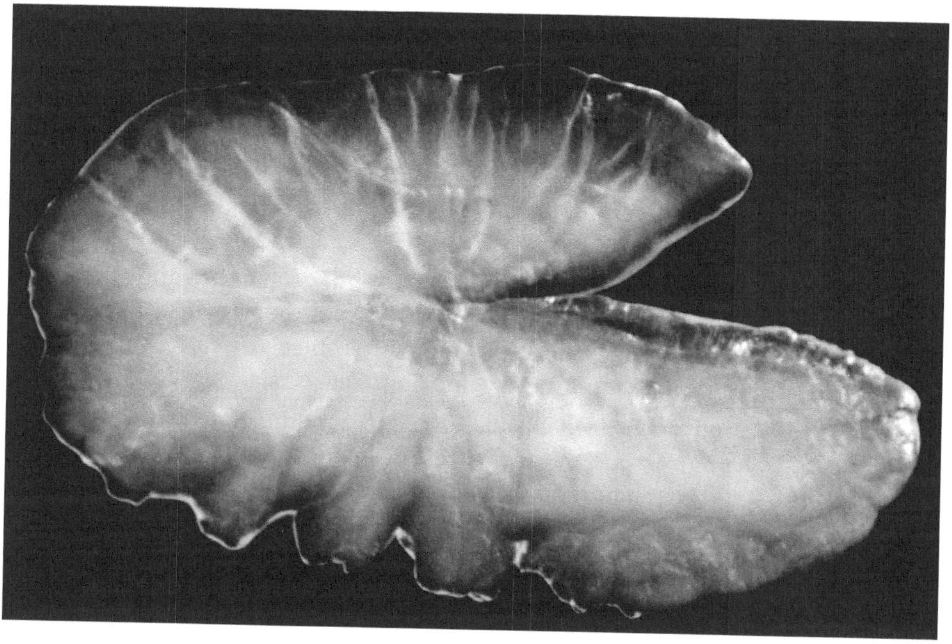

Plate 14: Otolith (sagitta) of *Gadusia chapra* (OL=3.10 mm; OH=1.82 mm) collected form the river Ganga.

Family– Clupidae

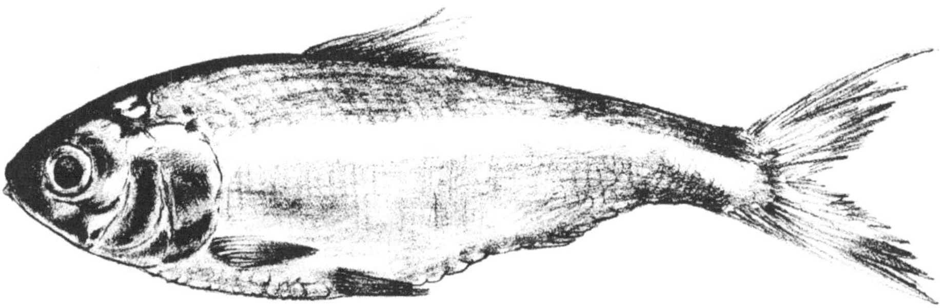

Plate 14a: *Gadusia chapra* collected form the river Ganga

Order: OSTEOGLOSSIFORMES

Family: Notopteridae (Featherbacks) - Plate 15 and Plate 15a

Species sampled	*Notopterus notopterus, Notopterus chitala*
Collection of samples	Gomti River, site: Lucknow (26.8467° N, 80.9462° E)
Habitat and Distribution (FishBase 2018)	Freshwaters of Asia: Indus, Ganges-Brahmaputra and Mahanadi river basin in India. However, members of the family are distributed in the freshwaters of Tropical Africa, Australia, New Guinea, South America and South east Asia.
General Characters (Jayaram 1999)	Moderately large fishes, elongate with the body broad, strongly compressed laterally, with fine scales on head also. Abdomen serrated before the pelvic fins. Branchiostegal rays eight or nine. No barbels. Dorsal fin single. Pelvic fins rudimentary or absent. Lateral line present.
Conservation status	*Notopterus notopterus* (Least Concern,) 2010, *Notopterus chitala* (Near Threatened, IUCN 2010)
Otolith morphology	Otolith shape: slightly rectangular to rhomboidal, Thickness: moderately thick, Form: Dorsal margin: flat with one hump in the middle, Ventral margin: slightly rounded and posteriorly sinuate. Posterior margin: sharply rounded and irregular, Rostrum: large, broad and rounded, Antirostrum: small and flat, Excisura: moderate and notch very shallow, angle acute.

Family-Notopteridae

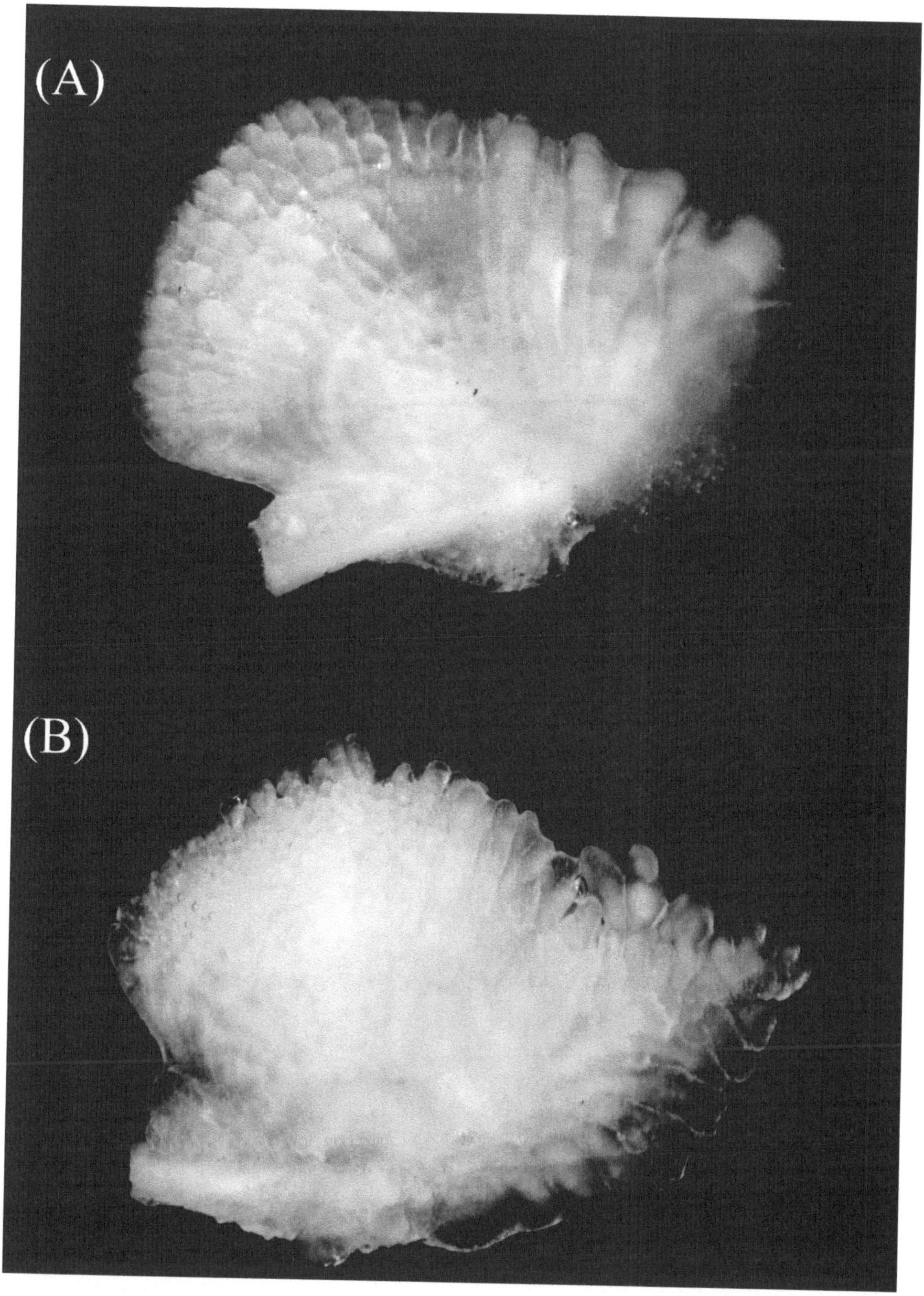

Plate 15: Otolith (Sagitta) of (A) *Notopterus notopterus* (OL=6.54 mm; OH=4.24 mm), (B) *Notopterus chitala* (OL=5.30 mm; OH=3.39 mm) collected form the river Gomti.

Family– Notopteridae

(A)

(B)

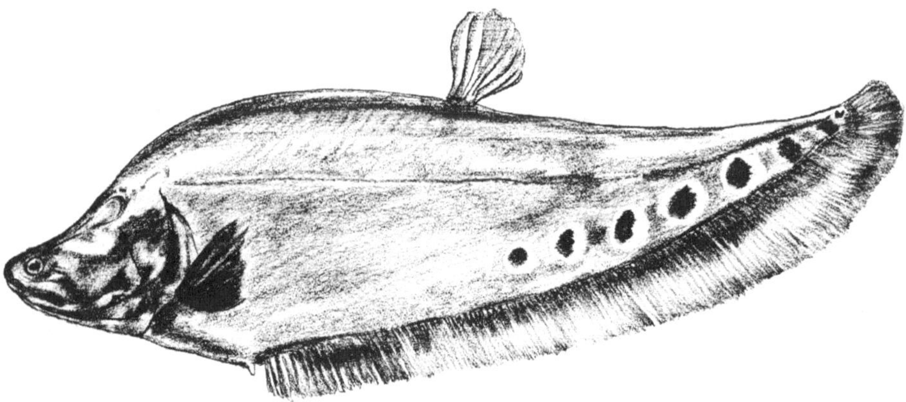

Plate 15a: (A) *Notopterus notopterus*, (B) *Notopterus chitala* collected form the river Gomti.

Order: MUGILIFORMES

Family: Mugilidae (Mullets) - Plate 16 and Plate 16a

Species sampled	*Sicamugil cascasia*
Collection of samples	Ganga River, site: Narora (28.1968° N, 78.3814° E)
Habitat and Distribution (FishBase)	Freshwaters of Asia: India, Pakistan and Bangladesh. However, members of the family are mostly marine, the economically valuable fishes are found often entering estuaries and tidal rivers.
General Characters (Jayaram 1999)	Body oblong to elongate, compressed, may be depressed anteriorly to a little extent. Head and body with scales. Eyes with or without adipose eyelids. Opercles with or without a spine. Teeth very fine or absent. Two dorsal fins short, widely separated, first with spines. Anal fin slightly longer than second dorsal fin. Pelvic fin sub-abdominal with a spine and suspended from an elongated shoulder bone. Lateral line absent.
Conservation status (IUCN)	Least concern
Otolith morphology	Otolith shape: oblong, Thickness: moderate, Dorsal margin: slightly rounded and irregular, Ventral margin: slightly rounded and irregular, Posterior margin: slightly rounded and irregular, Rostrum: short, broad and blunt, Antirostrum: minute and pointed. Excisura: moderate, notch very shallow, angle acute.

Family-Mugilidae

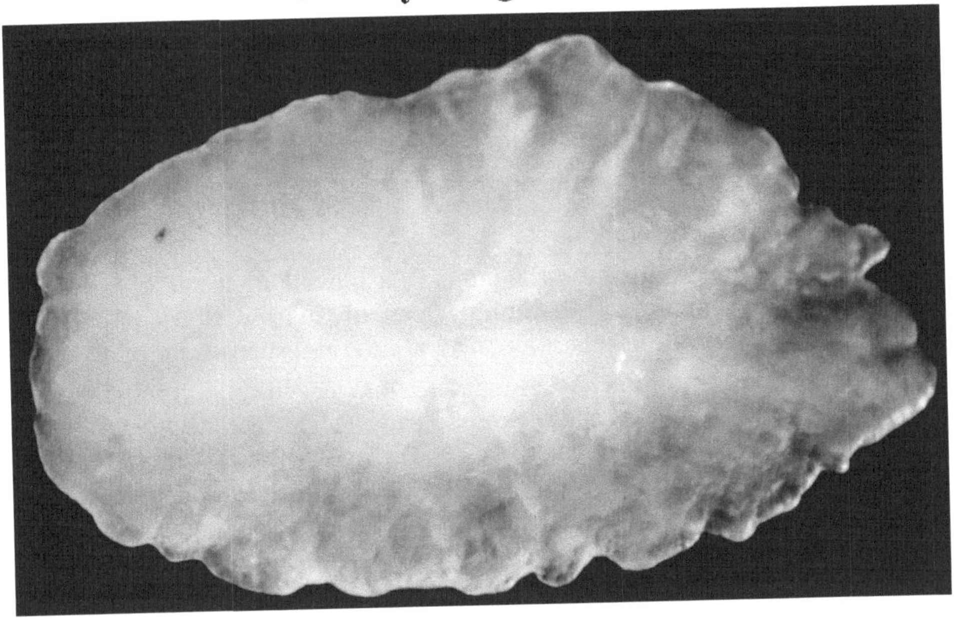

Plate 16: Otolith (sagitta) of *Sicamugil cascasia* (OL=2.77 mm; OH=1.55 mm) collected form the river Ganga.

Family– Muglidae

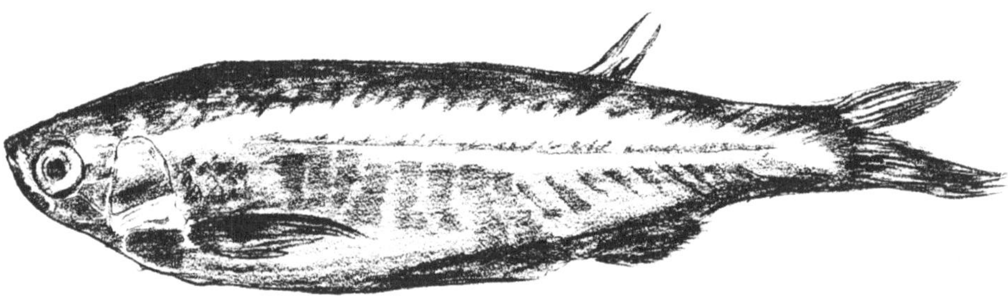

Plate 16a: *Sicamugil cascasia* collected form the river Ganga.

Order: CYPRINIFORMES

Family: Cyprinidae (Carps, Minnows)- Plate 17 and Plate 17a

Species sampled	*Labeo bata, Labeo boga, Labeo calbasu, Labeo victoriances, Labeo rohita, Catla catla, Puntius sarana, Puntius sophore, Puntius ticto, Osteobrama cotio, Salmostoma bacaila, Oxygaster gora, Cirrhinus mrigala, Cirrhinus reba, Cyprinus carpio, Ctenopharyngodon idella, Hypophthalmichthys nobilis, Crossocheilus latius*
Collection of samples	Ganga River, site: Narora (28.1968° N, 78.3814° E)
Habitat and Distribution (FishBase 2018)	Freshwaters of Asia: India, Pakistan, Nepal, Bangladesh, Myanmar, China. However, members of the family reported from Africa, Eurasia, North and Central America up to Southern Mexico. Also found in Australia, Madagascar, New Guinea and New Zealand.
General Characters (Jayaram 1999)	Otolith shape: ovate, Thickness: moderate, Dorsal margin: slightly rounded and irregular. Ventral margin: sharply rounded and irregular, Posterior margin: rounded and irregular, Rostrum: absent, Antirostrum: absent, Excisura: absent
Conservation status	*Labeo bata* (Least concern, IUCN 2011), *Labeo boga, Labeo calbasu, Labeo Rohita, Catla catla, Punctius sarana, Punctius ticto, Osteobrama cotio, Oxygaster gora, Cirrhinus mrigala,* (Least concern, IUCN 2010), *Labeo victoriances* (Critically Endangered, IUCN 2016), *Puntius sophore* (Not Evaluated), *Salmostoma bacaila* (Least concern, IUCN 2011), *Cirrhinus mrigala* (Least concern, IUCN 2010), *Cirrhinus reba* (Least concern, IUCN 2013), *Cyprinus carpio* (Vulnerable, IUCN 2008), *Ctenopharyngodon idella* (Not Evaluated), *Hypophthalmichthys nobilis* (Data Deficient, 2012), *Crossocheilus latius* (Least concern, IUCN 2018).
Otolith morphology	Otolith shape: Ovate, Thickness: Moderate, Form: Mesial approximately flat, Dorsal margin: Slightly rounded becoming flat and slightly irregular. Ventral margin: Slightly rounded and irregular, Posterior margin: Rounded, Rostrum: absent, Antirostrum: absent, Excisura: absent

Family-Cyprinidae

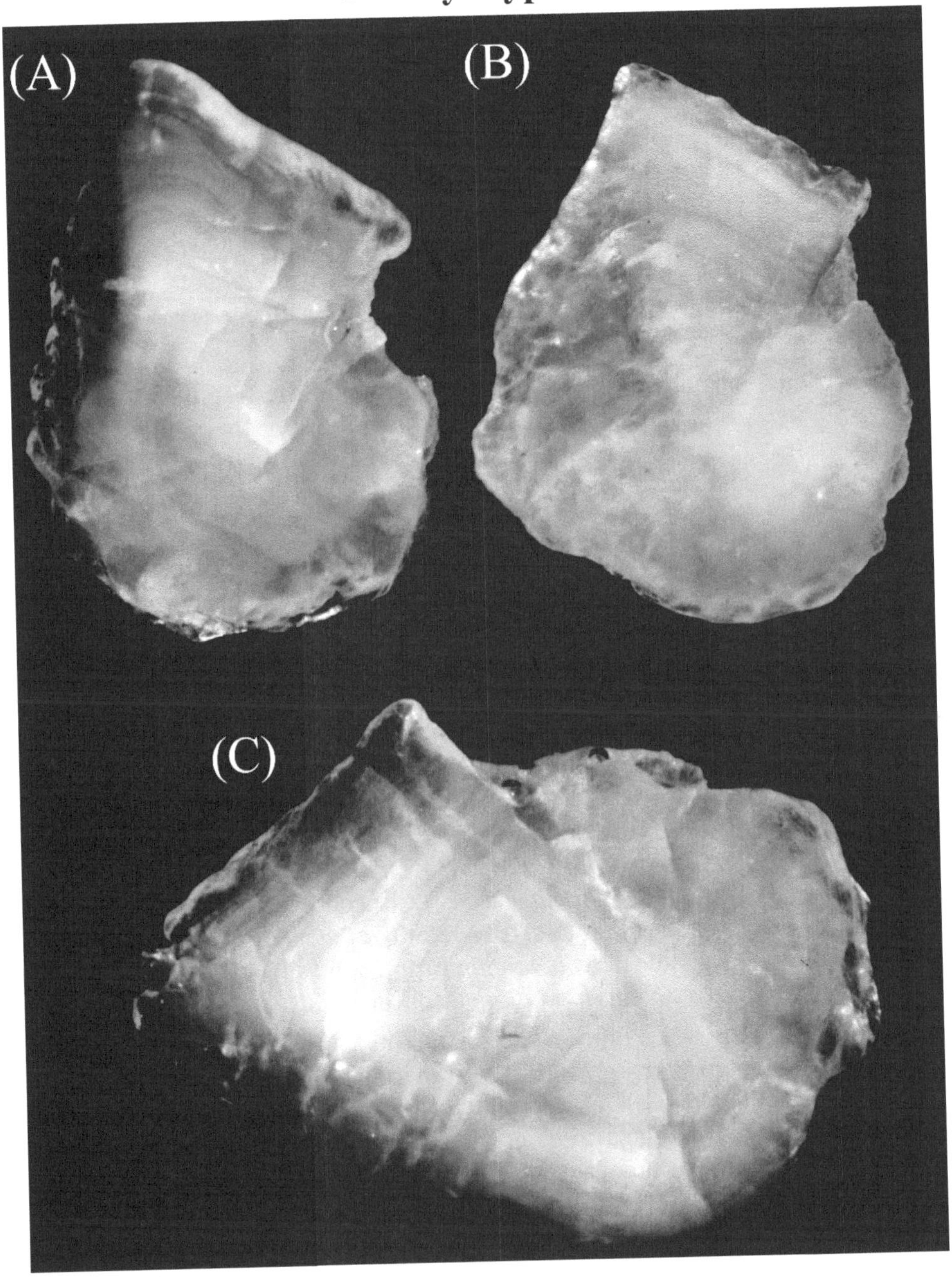

Plate 17: Otolith (asteriscus) of (A) *Labeo bata* (OL=2.48 mm; OH=1.68 mm), (B) *Labeo boga* (OL=1.69 mm; OH=1.57 mm), (C) *Labeo calbasu* (OL=2.39 mm; OH=1.45 mm) collected form the river Ganga.

Family-Cyprinidae

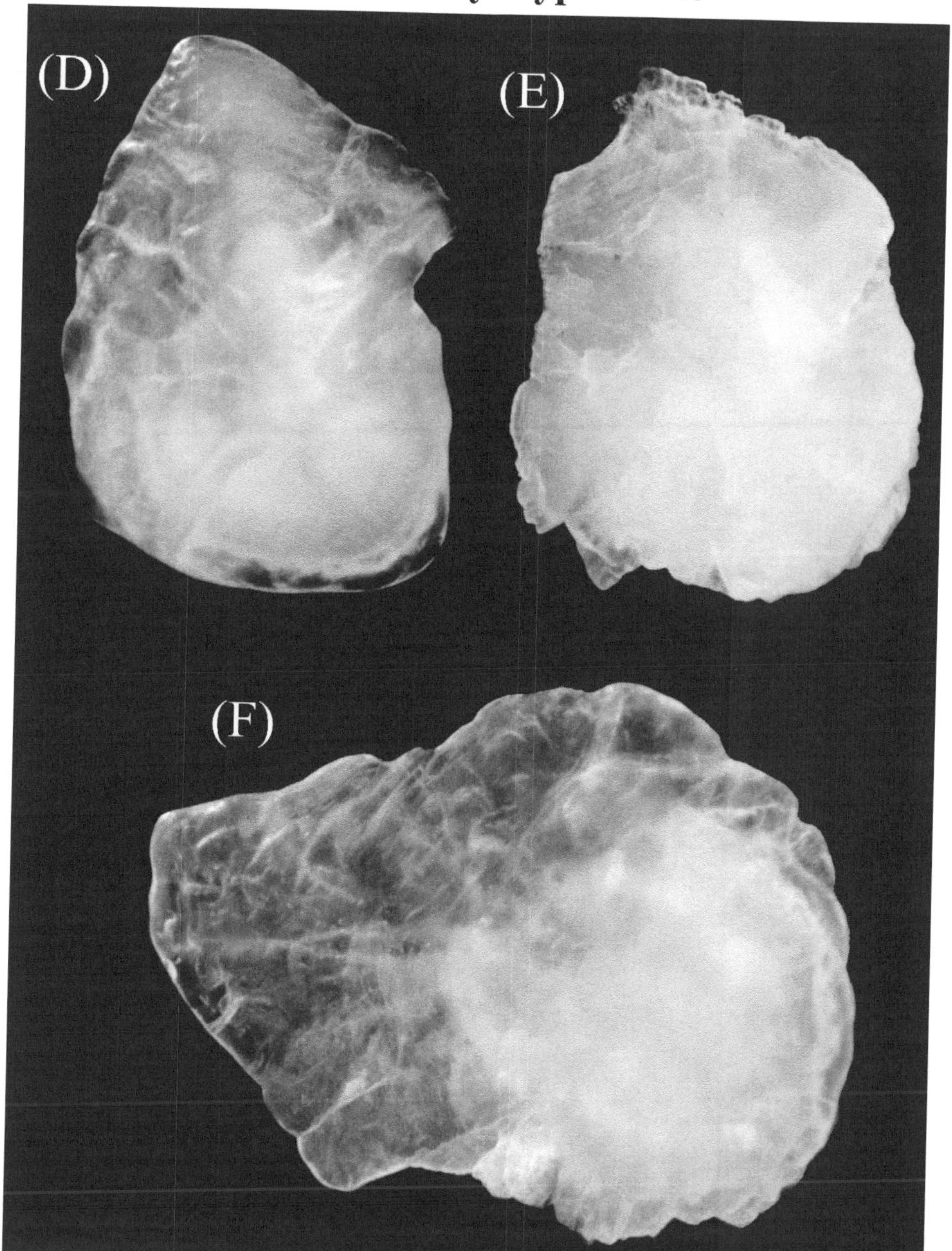

Plate 17: Otolith (asteriscus) of (D) *Labeo victoriances* (OL=1.58 mm; OH=1.12 mm), (E) *Labeo rohita* (OL=6.32 mm; OH=4.65 mm), (F) *Catla catla* (OL=4.12 mm; OH=3.84 mm) collected form the river Ganga.

Family-Cyprinidae

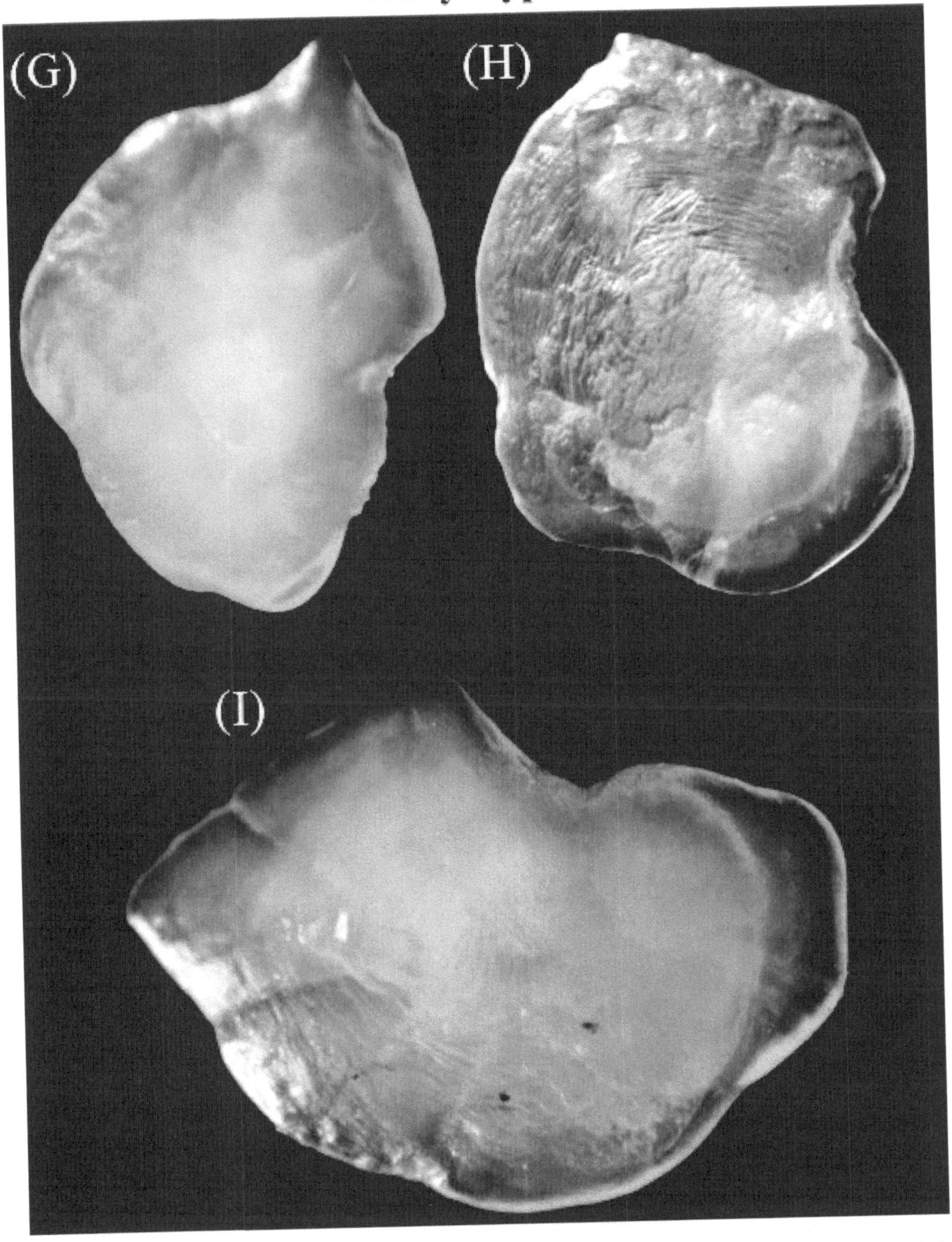

Plate 17: Otolith (asteriscus) of (G) *Puntius sarana* (OL=0.98 mm; OH=0.91 mm), (H) *Puntius sophore* (OL=1.10 mm; OH=0.86 mm), (I) *Puntius ticto* (OL=1.40 mm; OH=1.25 mm) collected form the river Ganga.

Family-Cyprinidae

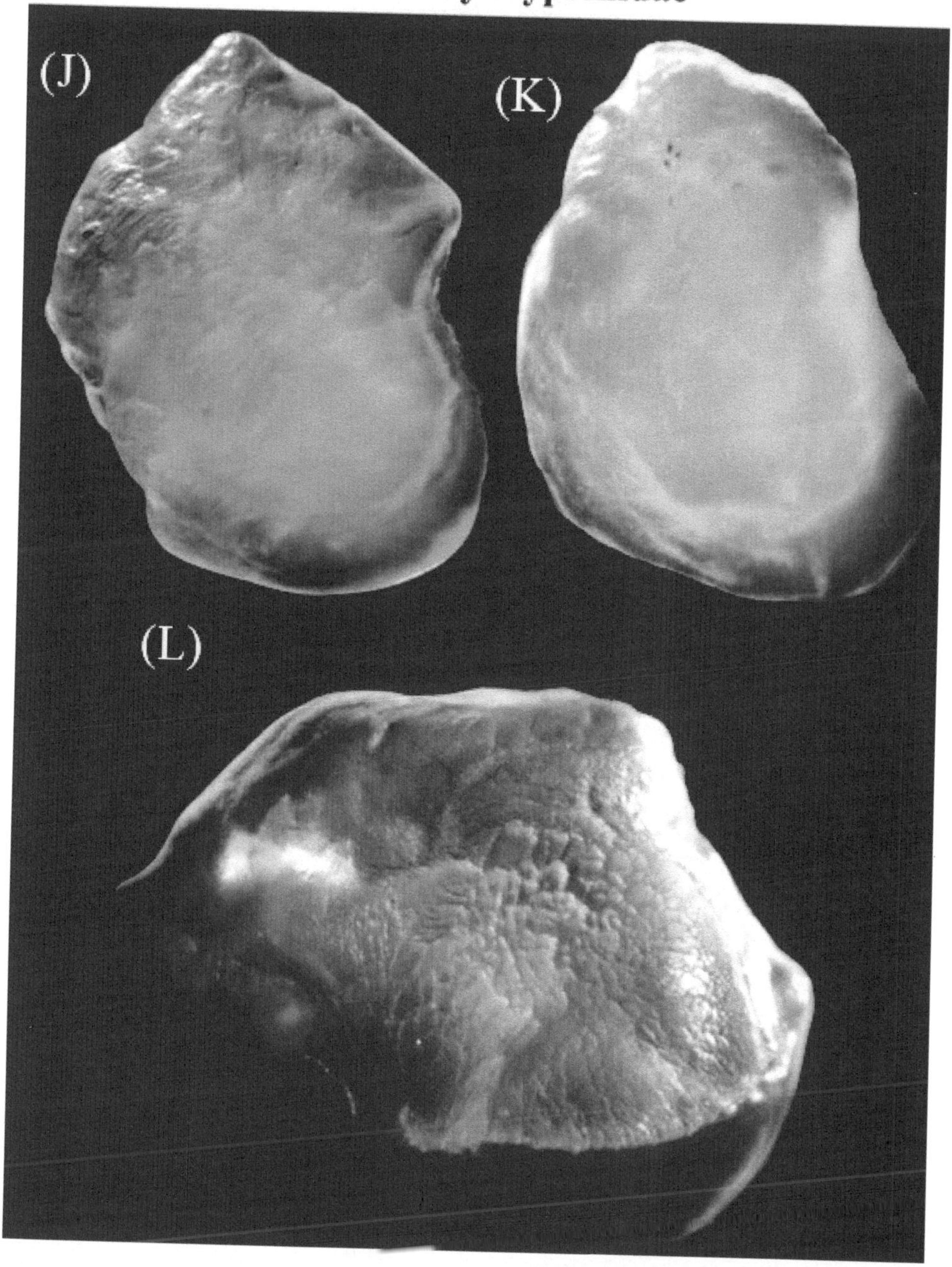

Plate 17: Otolith (asteriscus) of (J) *Osteobrama cotio* (OL=0.96 mm; OH=0.76 mm), (K) *Salmostoma bacaila* (OL=0.81 mm; OH=0.75 mm), (L) *Oxygaster gora* (OL=1.20 mm; OH=1.06 mm) collected form the river Ganga.

Family-Cyprinidae

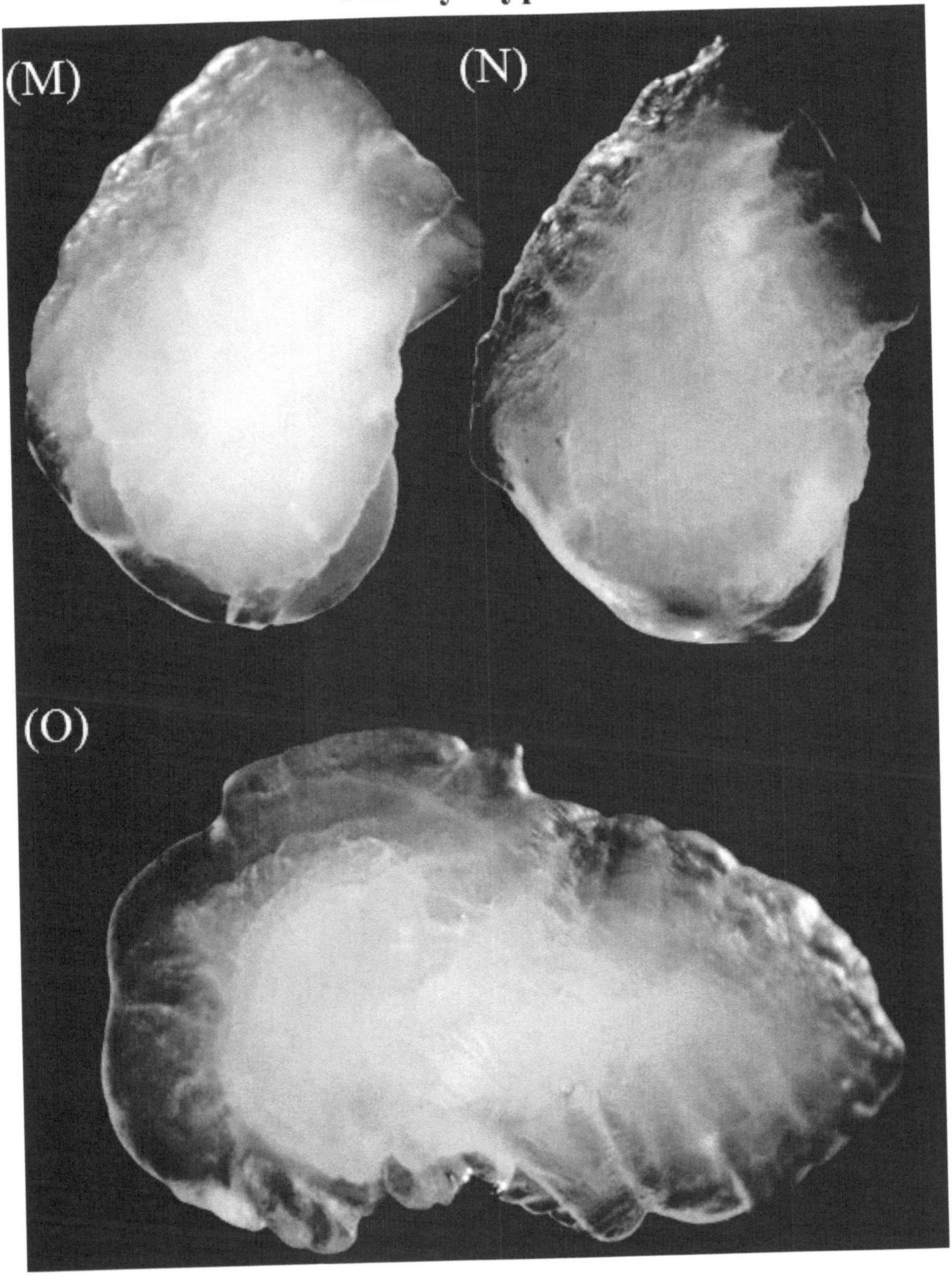

Plate 17: Otolith (asteriscus) of (M) *Cirrhinus mrigala* (OL=2.60 mm; OH=1.66 mm), (N) *Cirrhinus reba* (OL=2.21 mm; OH=1.68 mm), (O) *Cyprinus carpio* (OL=2.10 mm; OH=1.32 mm) collected form the river Ganga.

Family-Cyprinidae

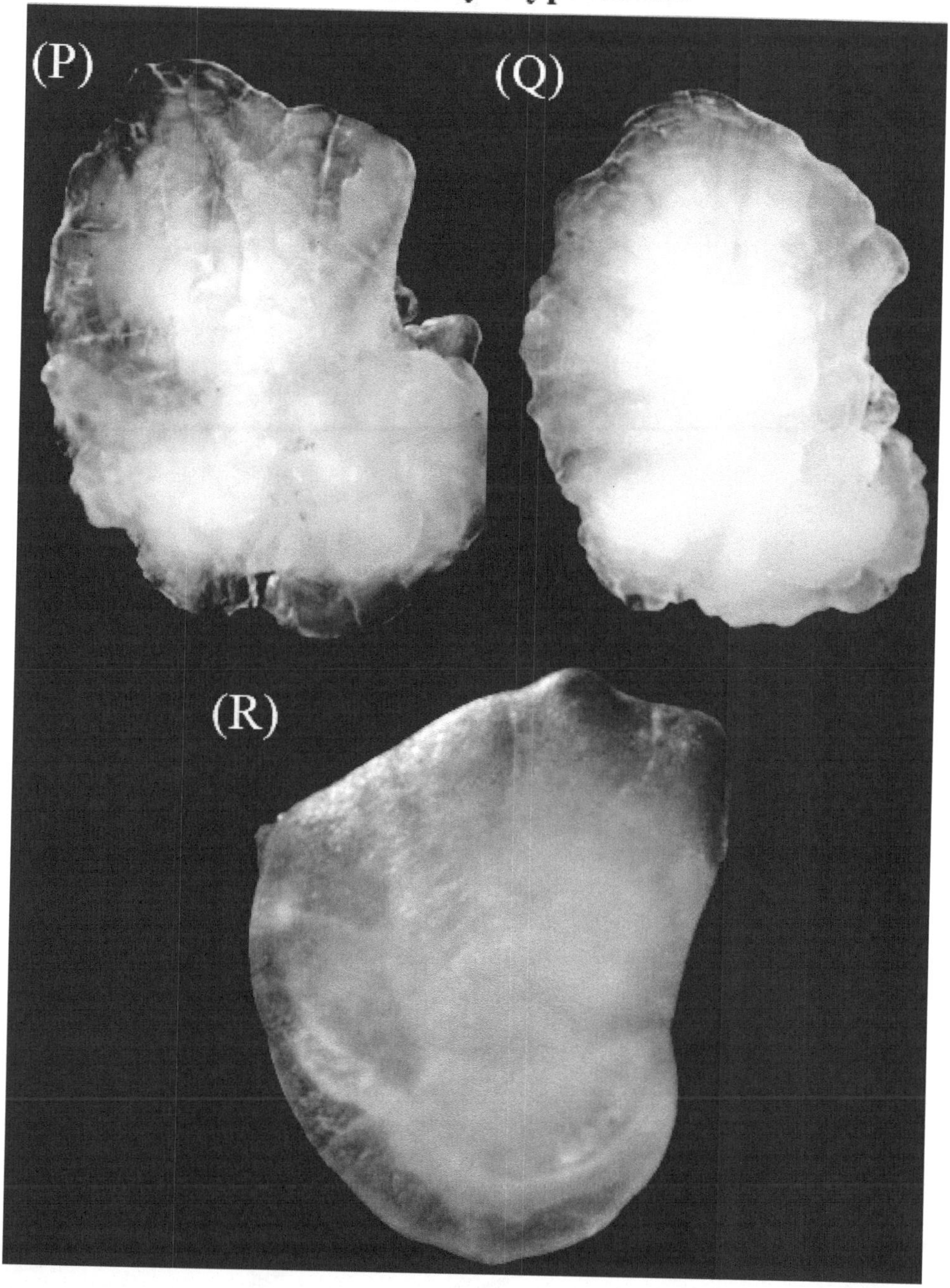

Plate 17: Otolith (asteriscus) of (P) *Ctenopharyngodon idella* (OL=2.17 mm; OH=1.77 mm), (Q) *Hypophthalmichthys nobilis* (OL=2.30 mm; OH=1.70 mm), (R) *Crossocheilus latius* (OL=1.12 mm; OH=1.02 mm) collected form the river Ganga.

Family– Cyprinidae

(A)

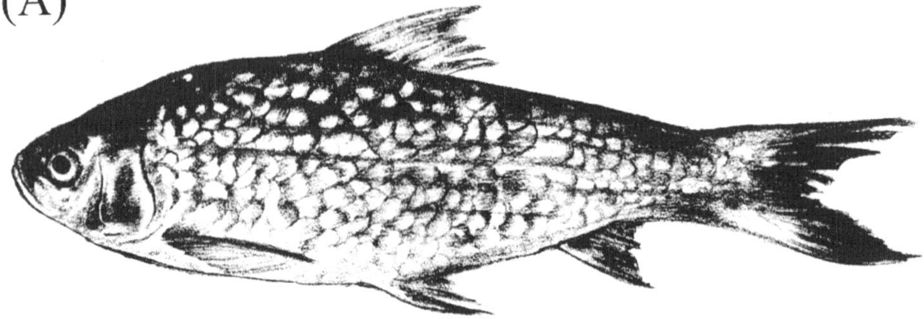

(B)

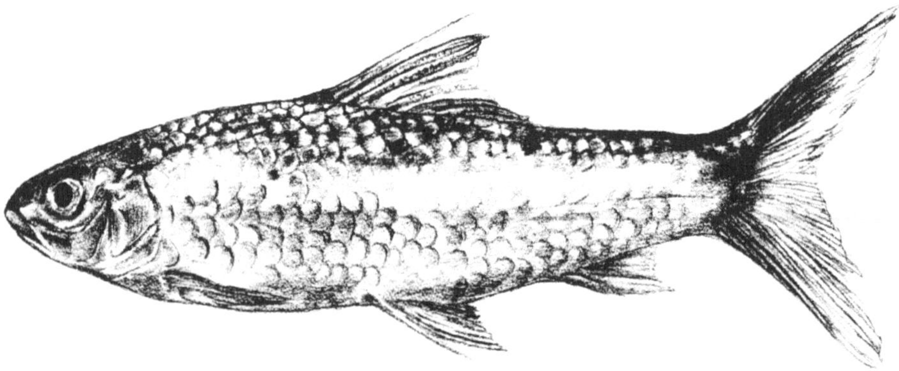

(C)

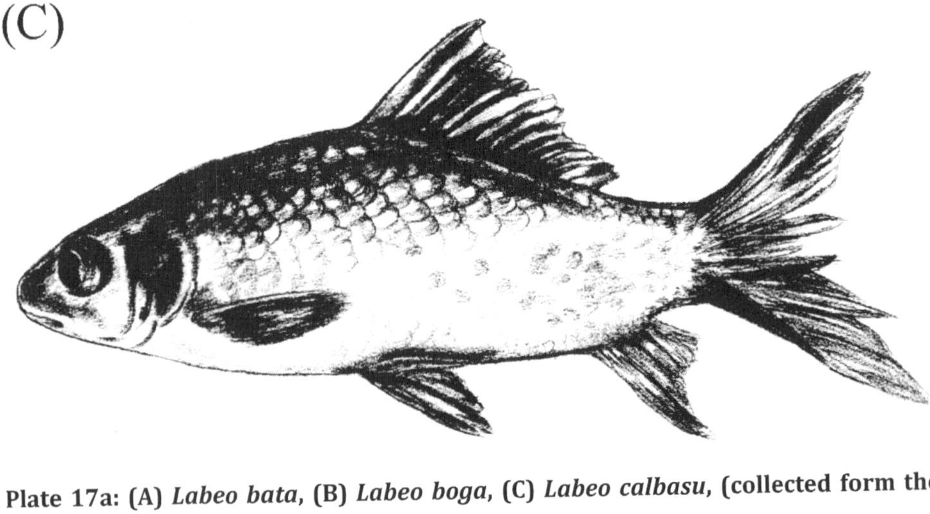

Plate 17a: (A) *Labeo bata*, (B) *Labeo boga*, (C) *Labeo calbasu*, (collected form the river Ganga.

Family– Cyprinidae

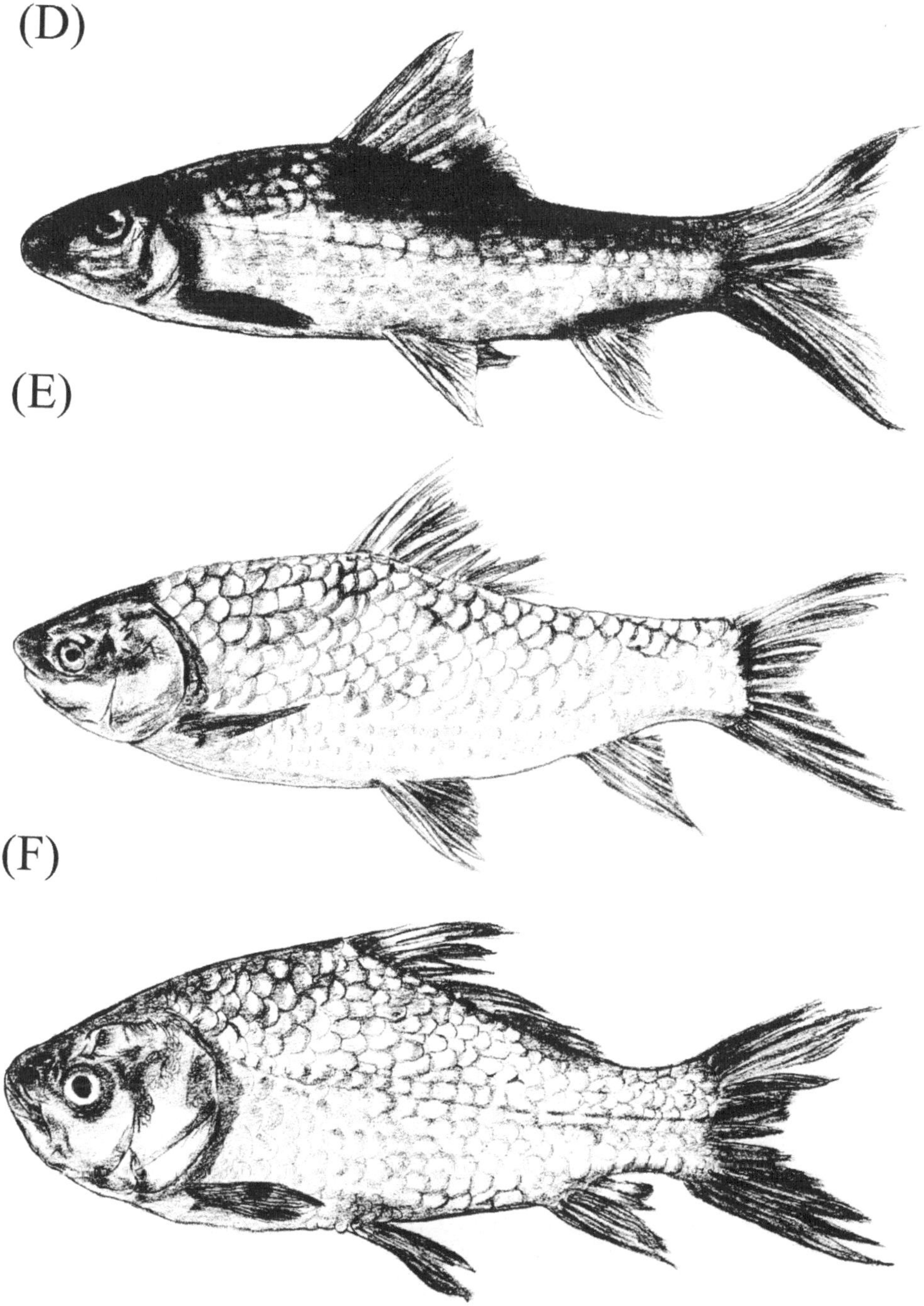

Plate 17a: (D) ***Labeo victoriances*****, (E)** ***Labeo rohita*****, (F)** ***Catla catla*** **collected form the river Ganga.**

Family– Cyprinidae

(G)

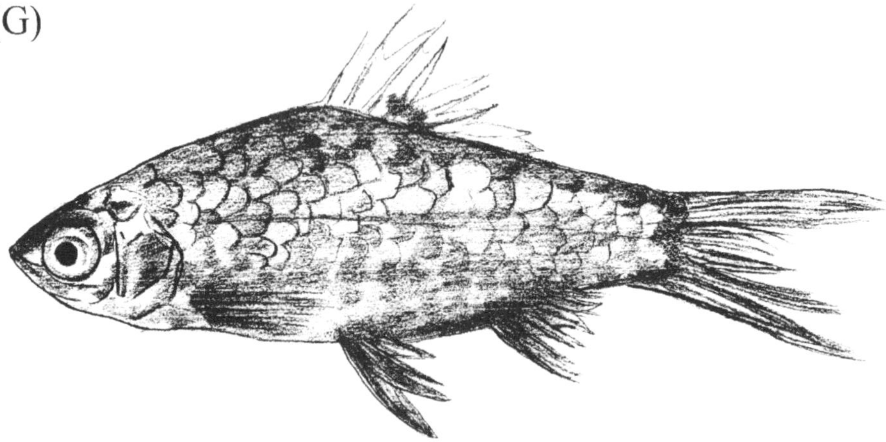

(H)

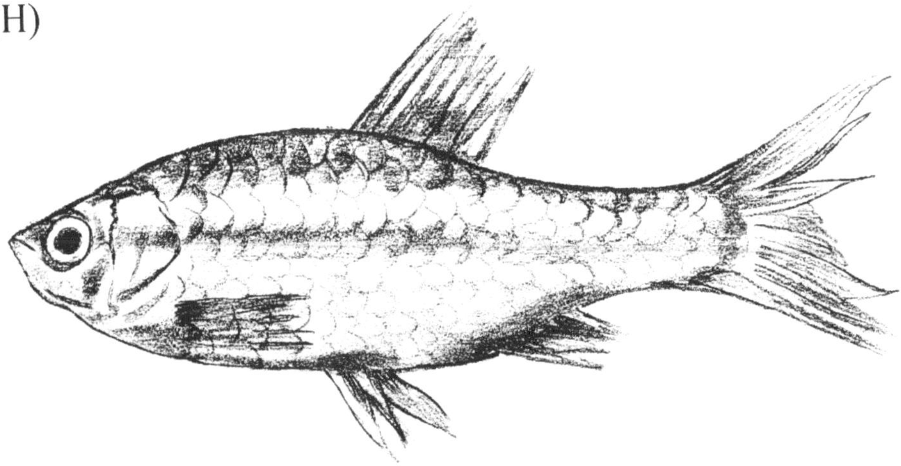

(I)

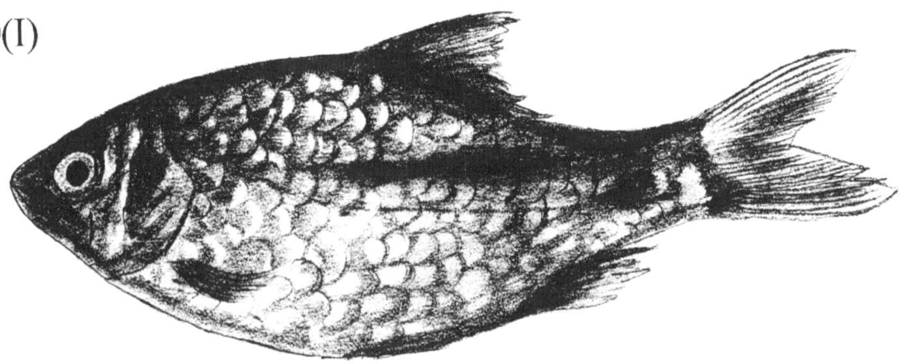

Plate 17a: (G) *Puntius sarana*, (H) *Puntius sophore*, (I) *Puntius ticto* collected form the river Ganga.

Family– Cyprinidae

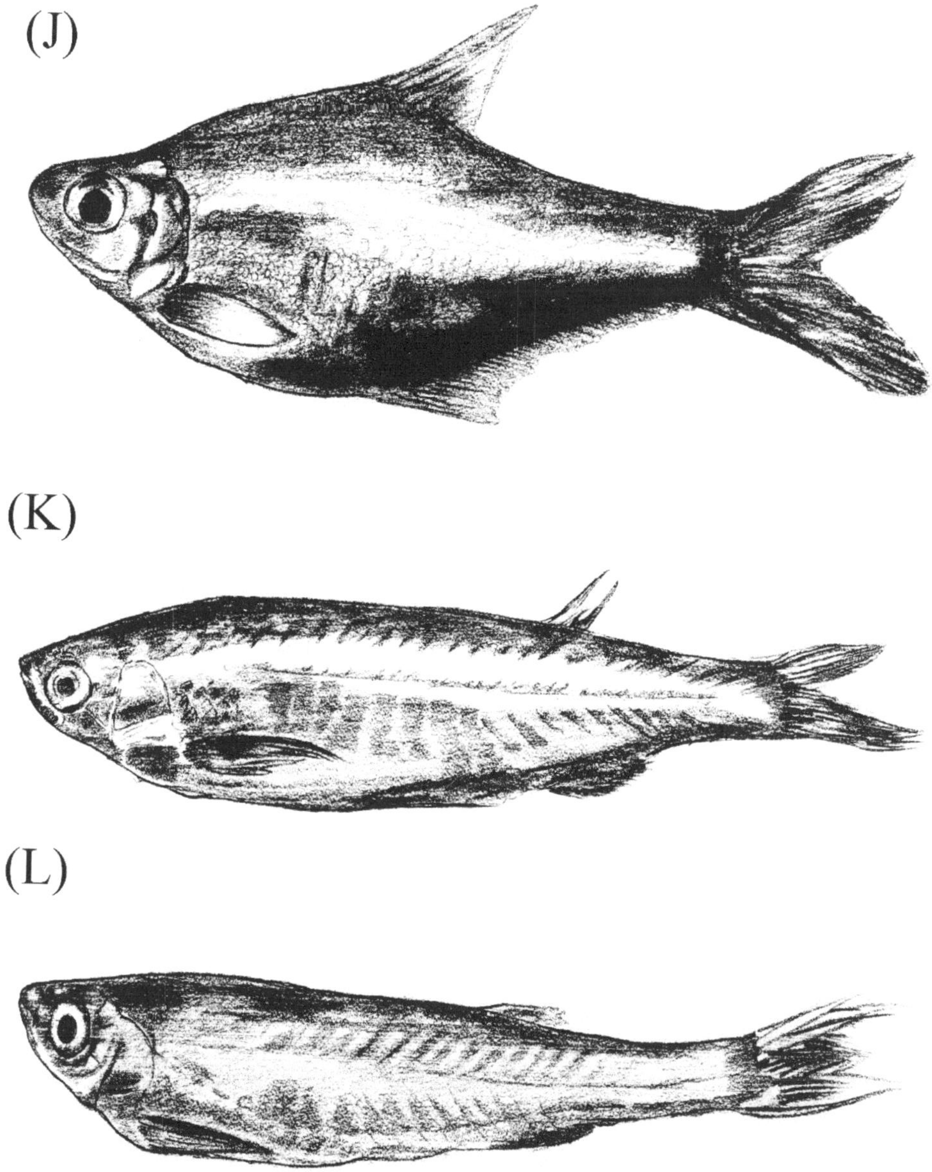

Plate 17a. (J) *Osteobrama cotio*, (K) *Salmostoma bacaila*, (L) *Oxygaster gora* collected form the river Ganga.

Family– Cyprinidae

(M)

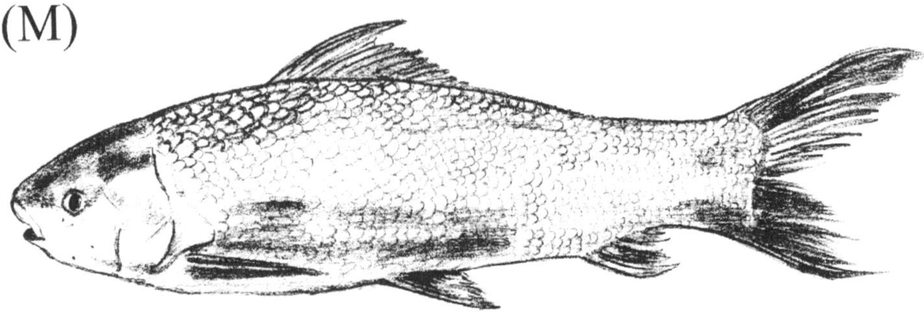

(N)

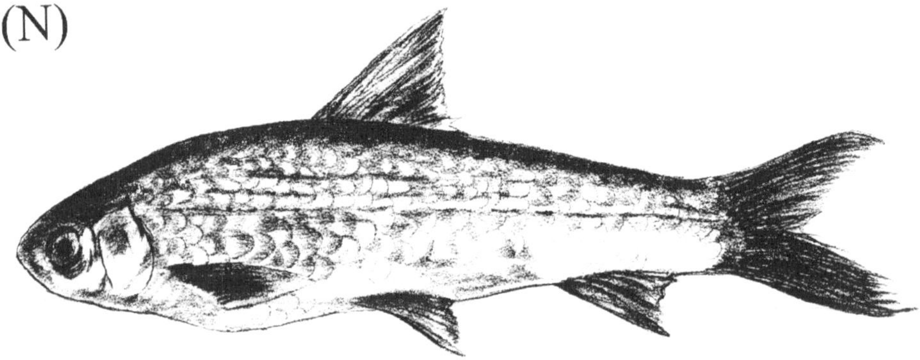

(O)

Plate 17a: (M) *Cirrhinus mrigala*, (N) *Cirrhinus reba*, (O) *Cyprinus carpio*, collected form the river Ganga.

Family– Cyprinidae

(P)

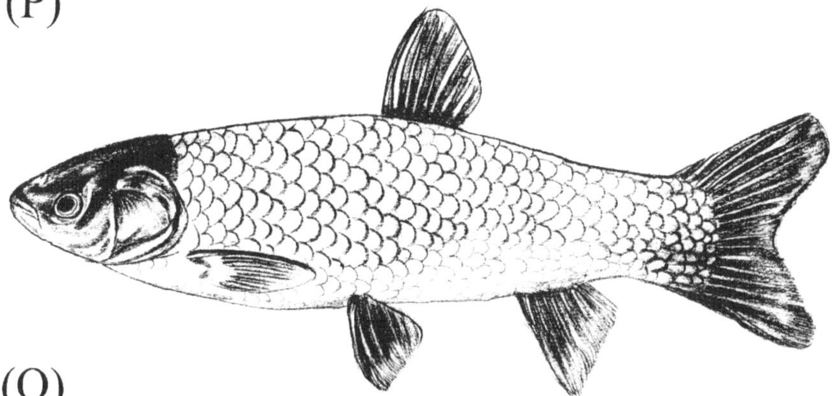

(Q)

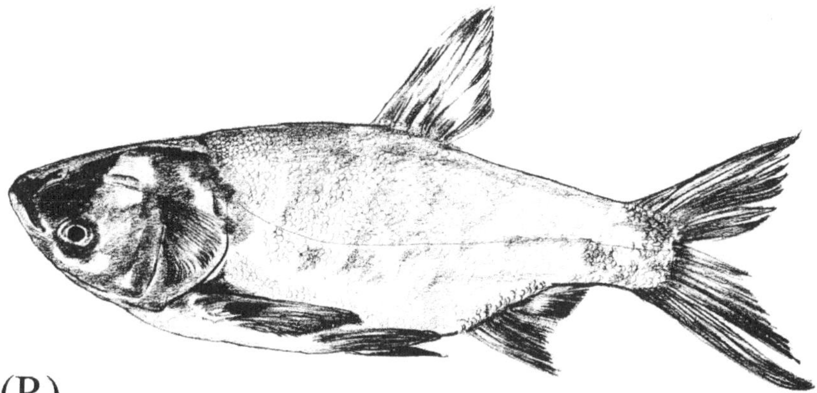

(R)

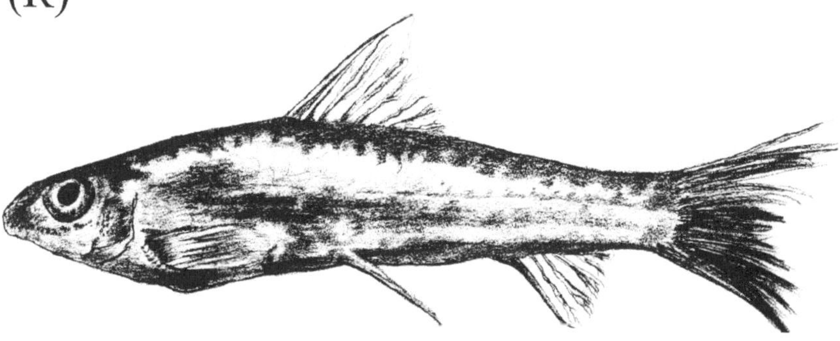

Plate 17a: (P) *Ctenopharyngodon idella*, (Q) *Hypophthalmichthys nobilis*, (R) *Crossocheilus latius* collected form the river Ganga.

References

Abilhôa, V., and Corrêa. M.F.M. (1992/93). Catálogo de otólitos de Carangidae (Osteichthyes – Perciformes) do Litoral do Estado do Paraná, Brasil. *Neritica* **7**, 119-131.

Aguirre, H., and Lombarte, A. (1999). Ecomorphological comparisons of sagittae in *Mullus barbatus* and *M. surmuletus*. *Journal of Fish Biology* **55**, 105-114.

Akkiran, N. (1985). A Systematic study on Carangidae (Pisces) Employing the otolith characters in the Eastern Mediterranean. *Bilješke-Notes* **63**: 1-9.

Allen, D.J. (2011). *Clarias batrachus*. The IUCN Red List of Threatened Species 2011: [Accessed on 20 July 2019].

Artzi, Y., Gisis, G., and Goldstein, H. (2009). Manual for freshwater fish otoliths-israel, Israel Nature and Parks Authority, Supported by The Fund for Nature Preservation, NPA.

Assis, C. A. (2000). Estudo morfológico dos otólitos Sagitta, Asteriscus e Lapillus de teleósteos (Actinopterygii, teleostei) de Portugal Continental: sua aplicação em estudos de filogenia, sistemática e ecologia. 1004 p. Tese (Doutorado em Biologia). Faculdade de Ciências - University of Lisbon.

ASSIS, C. A. (2005). The utricular otoliths, lapilli, of teleosts: the morphology and relevance for species identification and systematics studies. *Scientia Marina* **69**, 259-273.

Bremm, C.Q., and Schulz, U.H. (2014). Otolith atlas of fish of the Sinos River. *Brazilian Journal of Biology* **74,** 274-282.

Campana, S. E. (2004). Photographic atlas of fish otoliths of the Northwest Atlantic Ocean. Ottawa: NRC Research Press, p. 284 (Canadian Special Publication of Fisheries and Aquatic Sciences, 133).

Campana, S.E. (1999). Chemistry and composition of fish otoliths: pathways, mechanisms and applications. *Marine Ecology Progress Series* **188**, 263-297.

Casselman, J.M. (1987). Determination of age and growth. In The Biology of Fish Growth. Chap. 7. Edited by (A.H. Weatherley and H.S. Gill). pp. 209-242. (Academic Press, New York).

Campana, S.E., Thorrold, S.R., Jones, C.M., Gunther, D., Tubrett, M., Longerich, H., Jackson, S., Halden, N.M., Kalish, J.M., Piccoli, P., de Pontual, H., Troadec, H., Panfili, J., Secor, D.H., Severin, K.P., Sie, S.H., Thresher, R., Teesdale, W.J., Campbell, J.L. (1997) Comparison of accuracy, precision and sensitivity in elemental assays of fish otoliths using the electron microprobe, proton-induced X-ray emission, and laser ablation inductively coupled plasma mass spectrometry. *Canadian Journal of Fisheries and Aquatic Science.* **54**, 2068-2079

Chaudhry, S. (2010). *Channa gachua*. The IUCN Red List of Threatened Species 2010: e.T166123A6179047. [Accessed on 20 July 2019].

Chaudhry, S. (2010). *Channa marulius*. The IUCN Red List of Threatened Species 2010: [Accessed on 20 July 2019].

Chaudhry, S. (2010). *Channa punctata*. The IUCN Red List of Threatened Species 2010: [Accessed on 20 July 2019].

Chaudhry, S. (2010). *Channa striata*. The IUCN Red List of Threatened Species 2010: [Accessed on 20 July 2019].

Chaudhry, S. (2010). *Chitala chitala*. The IUCN Red List of Threatened Species 2010: [Accessed on 20 July 2019].

Chaudhry, S. (2010). *Gudusia chapra*. The IUCN Red List of Threatened Species 2010: [Accessed on 20 July 2019].

Chaudhry, S. (2010). *Securicula gora*. The IUCN Red List of Threatened Species 2010: [Accessed on 20 July 2019].

Chulin, A. K., and Chen, H. M. (2013). Comparative morphological study of otoliths in Taiwanese Anguilliformes fishes. *Journal of Marine Science and Technology* **21,** 77-85.

Cottrell, P. E., Trites, A. W. and Miller, E. H. (1996). Assessing the use of hard parts in faeces to identify harbour seal prey: results of captive feeding trials. *Canadian Journal of Zoology* **74**, 875-880.

Dahanukar, N. (2010). *Cirrhinus mrigala*. The IUCN Red List of Threatened Species 2010: [Accessed on 20 July 2019].

Dahanukar, N. (2010). *Labeo boga*. The IUCN Red List of Threatened Species 2010: [Accessed on 20 July 2019].

Dahanukar, N. (2010). *Labeo calbasu*. The IUCN Red List of Threatened Species 2010: [Accessed on 20 July 2019].

Dahanukar, N. (2010). *Labeo rohita*. The IUCN Red List of Threatened Species 2010: [Accessed on 20 July 2019].

Dahanukar, N. (2010). *Puntius sophore*. The IUCN Red List of Threatened Species 2010: [Accessed on 20 July 2019].

Dahanukar, N. (2010). *Systomus sarana*. The IUCN Red List of Threatened Species 2010: [Accessed on 20 July 2019].

Dahanukar, N. (2013). *Cirrhinus reba*. The IUCN Red List of Threatened Species 2013: [Accessed on 20 July 2019].

Dahanukar, N. (2018). *Minimugil cascasia* (amended version of 2010 assessment). The IUCN Red List of Threatened Species 2018: Available at [Accessed on 20 July 2019].

Dauphin, Y., and Dufour, E. (2003). Composition and properties of the soluble organic matrix of the otolith of a marine fish*: Gadus morhua Linne*, 1758 (Teleostei, Gadidae). *Comparative Biochemistry and Physiology. Part A, Molecular and Integrative Physiology* **134**, 551–561. doi:10.1016/S1095-6433(02)00358-6

Dey, S.C. (2010). *Xenentodon cancila*. The IUCN Red List of Threatened Species 2010: [Accessed on 20 July 2019].

Diouf, K., J. Panfili, M. Labonne, C. Aliaume, J. Tom´as, and T. D. Chi. (2006). Effects of salinity on strontium: calcium ratios in the otoliths of the West African black-chinned tilapia *Sarotherodon melanotheron* in a hypersaline estuary. *Environmental Biology of Fishes* **77**, 9–20.

Elsdon, T.S., Wells, B.K., Campana, S.E., Gillanders, B.M., Jones, C.M., Limburg, K.E.,Secor, D.H., Thorrold, S.R., Walther, B.D. (2008). Otolith chemistry to describe movements and life-history parameters of fishes: hypotheses, assumptions, limitations and inferencess. Oceanogr. *Oceanography and Marine Biology - An Annual Review* **46**, 297–330.

Enkins, A., Kullander, F.F. and Tan, H.H. (2009). *Johnius coitor*. The IUCN Red List of Threatened Species 2009: [Accessed on 20 July 2019].

Everson, I., Bendall, B. and Murray, A. (1999). Otolith and body size relationships in the mackerel icefish (*Champsocephalus gunnari*). *The convention on the conservation of Antarctic marine living resources* (*CCAMLR*) **6**, 117-123.

FishBase team RMCA and Geelhand, D. (2016). *Labeo victorianus*. The IUCN Red List of Threatened Species 2016: [Accessed on 20 July 2019].

Freyhof, J. and Kottelat, M. (2008). *Cyprinus carpio*. The IUCN Red List of Threatened Species 2008: [Accessed on 20 July 2018].

Freyhof, J., FishBase team RMCA and Geelhand, D. (2016). *Clarias gariepinus*. The IUCN Red List of Threatened Species 2016: [Accessed on 20 July 2018].

Froese, R. and D. Pauly. Editors. (2018). FishBase. World Wide Web electronic publication.

Frost, K. J. (1981). Descriptive key to the otoliths of gadid fishes of the Bering, Chukchi, and Beaufort Seas. *Artic* 34, 55-59.

Furlani, D., Gales, R., Pemberton, D. (2007). Otoliths of common Australian temperate fish: a photographic guide. p. 208. Collingwood: CSIRO Publishing.

Gomiero, L. M., and Braga, FMS. (2007). Descrição dos otólitos de tucunarés (Cichla sp. e *Cichla monoculus*) no reservatório da hidrelétrica de Volta Grande (SP-MG). *Ciência Animal Brasileira* **8**, 119-126.

Granadeiro, J. P. and Silva. M. A. (2000). The use of otoliths and vertebrae in the identification and size-estimation of fish in predator-prey studies. Cybium **24,** 383-393

Harvey, J.T., Loughlin, T.R., Perez, M.A., and Oxman, D.S. (2000). Relationship between Fish Size and Otolith Length for 63 Species of Fishes from the Eastern North Pacific Ocean. NOAA Technical Report NMFS 150. Washington, DC, USA: NOAA.

Hofstaetter, M., R.S Godefroid and P.H.B. Lemos. (1996). Morfologia dos otólitos de Umbrina coroides e Umbrina canosai (Pisces – Sciaenidae) do litoral do estado do Paraná, *Brazilian Archives of Biology and Technology* **39**, 157-164.

Huckstorf, V. (2012). *Hypophthalmichthys nobilis*. The IUCN Red List of Threatened Species 2012: [Accessed on 20 July 2019].

Hunt, J.J. (1992). Morphological characteristics of otolith for selected fish in the northwest Atlantic. *Journal of Northwest Atlantic Fishery Science*. **13**, 63-75.

Iizuka, K., and Katayama, S (2008): Otolith morphology of teleost fshes of Japan. *Bulletin of Fisheries Research Agency* **25**, 1–222

Jayaram, K. C. (1999). The Freshwater Fishes of the Indian Region. Narendra Publishing House, Delhi.

Jha, B.R., and Rayamajhi, A. (2010). *Heteropneustes fossilis*. The IUCN Red List of Threatened Species 2010: [Accessed on 20 July 2019].

Kalish, J. M. (1991). ^{13}C and ^{18}O isotopic disequilibria in fish otoliths: metabolic and kinetic effects. *Marine Ecology Progress Series* 75, 191–203. doi:10.3354/MEPS075191

Karen, Dwyer, S., Walsh, S. J. and Campana, S. E. (2003) Age determination, validation and growth of Grand Bank yellowtail flounder (*Limanda ferruginea*) *ICES Journal of Marine Science* **60**, 1123–1138. doi:10.1016/S1054–3139(03)00125-5

Khan, M. A., Khan, S. (2009). Comparison of age estimates from scale, opercular bone, otolith, vertebrae and dorsal fin ray in *Labeo rohita* (Hamilton), *Catla catla* (Hamilton) and *Channa marulius* (Hamilton). *Fisheries Research* **100,** 255-259.

Khan, M. A., Khan, S., and Khan, S. (2017). Precision of age estimates in striped snakehead, *Channa striata* (Bloch, 1793) from the Ganga River and its tributaries (rivers Gomti and Yamuna). *Journal of Applied Ichthyology* **33,** 230-235.

Khan, M. A., Khan, S., Ankita, Ahmed, I., and Nadeem, M., 2019. Age and growth of spotted snakehead, *Channa punctata* (Bloch, 1793) inhabiting the river Ganga. *Journal of Ichthyology,* 59, 197–204.

Khan, M.A., Miyan, K., Khan, S., Patel, D.K., Ansari, N.G. (2012). Studies on the elemental profile of otoliths and truss network analysis for stock discrimination of the threatened stinging catfish *Heteropneustes fossilis* (Bloch 1794) from the Ganga river and its tributaries. *Zoological Studies* **51**, 1195–1206.

Kumar, P., Chakraborty, S. K., and Jaiswar, A. K. (2012). Comparative otolith morphology of sciaenids occurring along the north-west coast of India. *Indian Journal of Fisheries* **59**, 19-27.

Larson, H., Britz, R. and Sparks, J.S. (2016). *Glossogobius giuris.* The IUCN Red List of Threatened Species 2016: [Accessed on 20 July 2019].

Limburg, K. E. (1995). Otolith strontium traces environmental history of subyearling American shad *Alosa sapidissima. Marine Ecology Progress Series* **119**, 25–35.

Limburg, K. E. (1998). Anomalous migration of anadromous herrings revealed with natural chemical tracers. *Canadian Journal of Fisheries and Aquatic Sciences* **55**, 431–437.

Lin, C.H. and Chang, C.W. (2012). Otolith atlas of Taiwan fishes. Published by National Museum of Marine Biology and Aquarium, Taiwan, 415 pp

Lombarte, A., and Cruz, A. (2007). Otolith size trends in marine fish communities from different depth strata. *Journal of Fish Biology* **71**, 53-76.

Lombarte, A., Chic, Ò., Parisi-Baradad, V., Olivella, R., Piera, J., García–Ladona, E. (2006). A web-based environment from shape analysis of fish otoliths: the AFORO database. *Scientia Marina* **70**, 147-152.

Lombarte, A., Palmer, M., Matallanas, J., Gómez-Zurita, J., Morales-Nin, B. (2010). Ecomorphological trends and phylogenetic inertia of otolith sagittae in Nototheniidae. *Environmental Biology of Fishes* **89**, 607-18.

Miyan, K., Khan, M.A., Patel, D.K., Khan. S, Ansari, N.G. (2016). Truss morphometry and otolith microchemistry reveal stock discrimination in *Clarias batrachus* (Linnaeus, 1758) inhabiting the Gangetic river system. *Fisheries Research* **173**, 294–302

Morrow, J. E. (1977). Illustrated keys to otoliths of forage fishes of the Gulf of Alaska, Bering Sea and Beaufort Sea. In: UNITED STATES. National Oceanic and Atmospheric Administration. Environmental assessment of the Alaskan Continental shelf: annual reports of principal investigators for the Year Ending

March 1977. Boulder: Outer Continental Shelf Environmental Assessment Program. p. 757-825.

Morrow, J. E. (1979). Preliminary keys to otoliths of some adult fishes of the Gulf of Alaska, Bering Sea and Beaufort Sea. Washington: U.S. Department of Commerce, NOAA, NMFS, p. 34 (NOAA Technical Report NMFS Circular, 420).

Ng, H.H. (2010) *Mystus cavasius*. The IUCN Red List of Threatened Species 2010: [Accessed on 20 July 2019].

Ng, H.H. (2010). *Bagarius bagarius*. The IUCN Red List of Threatened Species 2010: [Accessed on 20 July 2019].

Ng, H.H. (2010). *Clupisoma garua*. The IUCN Red List of Threatened Species 2010: [Accessed on 20 July 2019].

Ng, H.H. (2010). *Gagata cenia*. The IUCN Red List of Threatened Species 2010: [Accessed on 20 July 2019].

Ng, H.H. (2010). *Glyptothorax telchitta*. The IUCN Red List of Threatened Species 2010: [Accessed on 20 July 2019].

Ng, H.H. (2010). *Mystus tengara*. The IUCN Red List of Threatened Species 2010: [Accessed on 20 July 2019].

Ng, H.H. (2010). *Mystus vittatus*. The IUCN Red List of Threatened Species 2010: [Accessed on 20 July 2019].

Ng, H.H. (2010). *Nandus nandus*. The IUCN Red List of Threatened Species 2010: [Accessed on 20 July 2019].

Ng, H.H. (2010). *Notopterus notopterus*. The IUCN Red List of Threatened Species 2010: [Accessed on 20 July 2019].

Ng, H.H. (2010). *Rita rita*. The IUCN Red List of Threatened Species 2010: [Accessed on 20 July 2019].

Ng, H.H. (2010). *Sperata seenghala*. The IUCN Red List of Threatened Species 2010: [Accessed on 20 July 2019].

Ng, H.H. (2010). *Wallago attu*. The IUCN Red List of Threatened Species 2010: [Accessed on 20 July 2019].

Ng, H.H. and Dahanukar, N. (2011). *Ailia coila*. The IUCN Red List of Threatened Species 2011: [Accessed on 20 July 2019].

Nolf, D. (1979). Contribution a l'étude des otolithes des poissons. I. Morphologie comparée des otolithes (sagittae) des Dentex de la Méditerranée et de l'Atlantique Tropical africain. *Bulletin - Institut royal des sciences naturelles de Belgique* **51**, 1-13.

Nolf, D. (1985). Handbook of paleoichthyology: otolithi piscium. Stuttgart: Gustav Fischer Verlag, v. 10, p.145.

Nolf, D., and Bajpai, S. (1992). Marine Middle Eocene fish Otoliths from India and Java. *Bulletin De L'institut Royal Des Sciences Naturelles De Belgique, Sciences De La Terre* **62**, 195-221.

Panfili, J., De Pontual, H., Troadec, H., Wright, P.J., 2002. Manual of Fish Sclerochronology, Ifremer-IRD co-edition, 2002, 464 p.

Payan, P., De Pontual. H., Boeuf. G., and Mayer-Gostan, N. (2004). Endolymph chemistry and otolith growth in fish. *Comptes Rendus Palevol* **3**, 535-547.

Phelps, Q. E., Edwards, K. R., and Willis, D. W. (2007). Precision of five structures for estimating age of common carp. *North American Journal of Fisheries Management* **27**, 103–105.

Rana, R. S. (1988). Freshwater fish otoliths from the Deccan Trap associated sedimentary (Cretaceous-Tertiary transition) beds of Rangapur, Hyderabad, District, Andhra Pradesh, India, *Geobios* **21**, 465-493. https://doi.org/10.1016/S0016-6995(88)80045-7

Rayamajhi, A., Jha, B.R., and Sharma, C.M. (2010). *Mastacembelus armatus*. The IUCN Red List of Threatened Species 2010: [Accessed on 20 July 2019].

Rema Devi, K.R. and Ali, A. (2011). *Labeo bata*. The IUCN Red List of Threatened Species 2011: [Accessed on 20 July 2019].

Rema Devi, K.R. and Dahanukar, N. (2011). *Salmophasia bacaila*. The IUCN Red List of Threatened Species 2011: [Accessed on 20 July 2019].

Rema Devi, K.R. and Raghavan, R. (2011) *Sperata aor*. The IUCN Red List of Threatened Species 2011: [Accessed on 20 July 2019].

Rojo, A.L. (2015). Report in the Fish Otolith Collection at the Nova Scotia Museum, Curatorial Report Number 105, Nova Scotia Museum, Halifax.

Rousset, J. (1983). Etude des ecailles et otolithes des soleides d'Algerie. *Cybium* **7**, 71-96.

Sánchez, R. O., and Martínez, V. H. (2017). Morphological variations of the three otoliths of some species of the family Loricariidae (Ostariophysi: Siluriformes), *Neotropical Ichthyology* 15(1): e160058. *DOI: 10.1590/1982-0224-20160058*

Secor, D. H., and J. R. Rooker. (2000). Is otolith strontium a useful scalar of life cycles in estuarine fishes? *Fisheries Research* **46**, 359–371.

Shaji, C.P. (2011). *Lepidocephalus thermalis*. The IUCN Red List of Threatened Species 2011: [Accessed on 20 July 2019].

Singh, L.K. (2018). *Tariqilabeo latius* (amended version of 2010 assessment). The IUCN Red List of Threatened Species 2018: [Accessed on 20 July 2019].

Sipe, A. M., and Chittenden, Jr. M. E. (2001). A comparison of calcified structures for aging summer flounder, *Paralichthys dentatus*. *Fishery Bulletin*. **99**, 628–640.

Smale, M. J., Watson, G., and Flecht, T. (1995). Otolith atlas of southern African marine fishes. Ichthyological Monographs of the J.L.B. Smith Institute of Ichthyology, No. 1, xiv, 253 pages, 149 plates.

Tenzin, K. (2010). *Gibelion catla*. The IUCN Red List of Threatened Species 2010: [Accessed on 20 July 2019].

Tollit, D. J., Steward, M. J., Thompson, P. M., Pierce, G. J., Santos, M. B. and Hughes, S. (1997). Species and size differences in the digestion of otoliths and beaks: implications for estimates of pinniped diet composition. *Canadian Journal of Fisheries and Aquatic Sciences* **54**, 105-1197.

Torno, M.F. (1976). Descripción e comparación de los otolitos de algunas familias de peces de la plataforma Argentina. Revista del Museo Argentino de Ciencias Naturales. *Bernardino Rivadavia* **12**, 25-44.

Tuset, V. M., Lombarte, A., Assis, C. A. (2008). Otolith atlas for the Western Mediterranean, North and Central Eastern Atlantic. *Scientia Marina* **72S1**, 7-198.

Tuset, V. M., Rosin, P. L., Lombarte, A. (2006): Sagittal otolith shape used in the identification of fishes of the genus Serranus. *Fisheries Research* **81**, 16-325.

Vidthayanon, C. (2012). *Parambassis ranga*. The IUCN Red List of Threatened Species 2012: [Accessed on 20 July 2019].

Vishwanath, W. (2010). *Macrognathus pancalus*. The IUCN Red List of Threatened Species 2010: [Accessed on 20 July 2019].

Vishwanath, W. (2010). *Osteobrama cotio*. The IUCN Red List of Threatened Species 2010: [Accessed on 20 July 2019].

Volpedo, A. V., and Echeverría, D. D. (2000). Catálogo y claves de otolitos para la identificación de peces del Mar Argentino. Buenos Aires: Editorial Dunken, 90 p. [v. 1. Peces de importância económica].

Wirtz, P. (1976). The otoliths of the Mediterranean Tripterygion. *Vie Milieu* **26**, 293-298. www.fishbase.org, version (06/2018).

Zimmerman, C. E. (2005). Relationship of otolith strontium-to-calcium ratios and salinity: experimental validation for juvenile salmonids. *Canadian Journal of Fisheries and Aquatic Sciences* **62**, 88–97.

Appendix

List of the Fishes and Shape of their Otoliths Included in the Book

Fish Name	***Family***	***Location***	***Otolith Shape***
Channa marulius	Channidae	Gomti river	Approximately oval
Channa punctata	Channidae	Gomti river	Approximately oval
Channa striata	Channidae	Gomti river	Approximately oval
Channa gachua	Channidae	Gomti river	Approximately oval
Glossogobius giuris	Gobiidae	Ganga river	Approximately square
Nandus nandus	Nandidae	Yamuna river	Oblong to fusiform
Parambassis ranga	Ambassidae	Ganga river	Slightly ovate
Ompok pabda	Siluridae	Ganga river	Oval
Ailia coila	Schilbeidae	Gomti river	Ovate to rectangular
Clupisoma garua	Schilbeidae	Gomti river	Ovate to rectangular
Clarias batrachus	Claridae	Yamuna river	Square, lobed and serrated
Clarias gariepinus	Claridae	Yamuna river	Square, lobed and serrated
Gagata cenia	Sisoridae	Gomti river	Slightly triangular
Glyptothorax telchitta	Sisoridae	Gomti river	Slightly triangular

Fish Name	***Family***	***Location***	***Otolith Shape***
Bagarius bagarius	Sisoridae	Gomti river	Slightly triangular
Heteropneustus fossilis	Heteropneustidae	Yamuna river	Rectangular and irregular
Sperata aor	Bagridae	Gomti river	Oval to irregular oblong
Mystus cavasius	Bagridae	Gomti river	Oval to irregular oblong
Mystus tengra	Bagridae	Gomti river	Oval to irregular oblong
Sperata seenghala	Bagridae	Gomti river	Oval to irregular oblong
Mystus vittatus	Bagridae	Gomti river	Oval to irregular oblong
Rita rita	Bagridae	Gomti river	Oval to irregular oblong
Wallago attu	Siluridae	Ganga river	Rectangular and irregular
Botia lohachata	Cobitidae	Ganga river	Approximately circular
Lepidocephalus thermalis	Cobitidae	Ganga river	Approximately circular
Mastacembelus armatus	Mastacembelidae	Ganga river	Oblong
Macrognathus pancalus	Mastacembelidae	Ganga river	Oblong
Macrognathus aculeatus	Mastacembelidae	Ganga river	Oblong
Xenentodon cancila	Belonidae	Gomti river	Nearly oblong
Gadusia chapra	Clupidae	Ganga river	Ovate
Notopterus notopterus	Notopteridae	Gomti river	Slightly rectangular to rhomboidal
Notopterus chitala	Notopteridae	Gomti river	Slightly rectangular to rhomboidal
Sicamugil cascasia	Mugilidae	Ganga river	Oblong
Labeo bata	Cyprinidae	Ganga river	Ovate
Labeo boga	Cyprinidae	Ganga river	Ovate
Labeo calbasu	Cyprinidae	Ganga river	Ovate

Fish Name	Family	Location	Otolith Shape
Labeo victoriances	Cyprinidae	Ganga river	Ovate
Labeo rohita	Cyprinidae	Ganga river	Ovate
Catla catla	Cyprinidae	Ganga river	Ovate
Puntius sarana	Cyprinidae	Ganga river	Ovate
Puntius ticto	Cyprinidae	Ganga river	Ovate
Puntius sophore	Cyprinidae	Ganga river	Ovate
Osteobrama cotio	Cyprinidae	Ganga river	Ovate
Salmostoma bacaila	Cyprinidae	Ganga river	Ovate
Oxygaster gora	Cyprinidae	Ganga river	Ovate
Cirrhinus mrigala	Cyprinidae	Ganga river	Ovate
Cirrhinus reba	Cyprinidae	Ganga river	Ovate
Cyprinus carpio	Cyprinidae	Ganga river	Ovate
Ctenopharyngodon idella	Cyprinidae	Ganga river	Ovate
Hypophthalmichthys nobilis	Cyprinidae	Ganga river	Ovate
Crossocheilus latius	Cyprinidae	Ganga river	Ovate

Index

G

H

I

L

M

N

O

P

R

S

W

X

Y

9 789354 614767

Printed at Repro India Ltd.